OECD e-Government Studies

Reaping the Benefits of ICTs in Spain

STRATEGIC STUDY ON COMMUNICATION INFRASTRUCTURE AND PAPERLESS ADMINISTRATION

This work is published on the responsibility of the Secretary-General of the OECD. The opinions expressed and arguments employed herein do not necessarily reflect the official views of the Organisation or of the governments of its member countries.

This document and any map included herein are without prejudice to the status of or sovereignty over any territory, to the delimitation of international frontiers and boundaries and to the name of any territory, city or area.

ISBN 978-92-64-11060-1 (print)
ISBN 978-92-64-17322-4 (PDF)
http://dx.doi.org/10.1787/9789264173224-en

Series: OECD e-Government Studies
ISSN 1990-1062 (print)
ISSN 1990-1054 (online)

The statistical data for Israel are supplied by and under the responsibility of the relevant Israeli authorities. The use of such data by the OECD is without prejudice to the status of the Golan Heights, East Jerusalem and Israeli settlements in the West Bank under the terms of international law.

Photo credits: Cover © Franck Boston/Shutterstock.com.

Corrigenda to OECD publications may be found on line at: *www.oecd.org/publishing/corrigenda*.
© OECD 2013

You can copy, download or print OECD content for your own use, and you can include excerpts from OECD publications, databases and multimedia products in your own documents, presentations, blogs, websites and teaching materials, provided that suitable acknowledgement of OECD as source and copyright owner is given. All requests for public or commercial use and translation rights should be submitted to *rights@oecd.org*. Requests for permission to photocopy portions of this material for public or commercial use shall be addressed directly to the Copyright Clearance Center (CCC) at *info@copyright.com* or the Centre français d'exploitation du droit de copie (CFC) at *contact@cfcopies.com*.

Foreword

This report contains the main findings of a strategic review on communication infrastructure and paperless administration in Spain, the *Plan Avanza 2* (hereinafter, "the Plan"*)*. The OECD reviewed in 2010 the previous information society strategy, *Plan Avanza*, and published the report *Good Governance for Digital Policies: How to get the Most Out of ICT. The Case of Spain's Plan Avanza.* Spain launched *Plan Avanza* in 2005 and, to sustain its efforts to advance its information society, approved, in July 2010, *Plan Avanza 2*. The recently appointed Spanish government is in the process of elaborating the Digital Agenda for Spain, aligned with the European Digital Agenda.

Addressing the major challenges currently faced by Spain, this information society strategy has aimed to foster economic growth and to place the country in a leadership position in the advanced use of ICT services and products to improve public service delivery and overall national competitiveness. This review provides ground to further improve these policies and their outcomes.

This strategic study focuses on two key areas: improving communication infrastructure and achieving a paperless administration through e-government. The analysis of e-government contained in this report is based on the OECD analytical framework developed on the basis of e-government country reviews and of the following reports: *The e-Government Imperative* (2003), *e-Government for Better Government* (2005), *Rethinking e-Government Services: User-centred Approaches* (2009), and *M-Government: Mobile Technologies for Responsive Governments and Connected Societies* (2011). The review was carried out under the auspices of the OECD Network on E-Government, which considered its main findings as part of the Programme of Work of the Public Governance and Territorial Development Directorate.

Undertaken at the request of the Spanish government in 2011, the review was conducted under the previous government administration and does not reflect changes in the composition of the government and its ministries made after December 2011. Under the new government several ministries were restructured. Among them, the Ministry of Industry, Tourism and Trade became the Ministry of Industry, Energy and Tourism, in which the Secretariat of State for Telecommunications and the Information Society remains.

The report was conceived by the Directorate for Public Governance and Territorial Development and the Directorate for Science, Technology and Innovation in collaboration with the government of Spain. Chapter 1 was drafted by Tony Shorthall (Telage) and Agustín Díaz-Pinés under the direction of Dimitri Ypsilanti. Chapter 2 was written by Adam Mollerup, under the direction of János Bertók and Barbara Chiara Ubaldi. The introduction and the executive summary were jointly drafted by the authors. Administrative assistance was provided by Karine Ravet and Sarah Michelson. Lia Beyeler, Karena Garnier and Simon Gregg assisted in the editing of the report and Jennifer Allain prepared the manuscript for publication.

The OECD would like to thank the Spanish administration, especially the Secretariat of State for Telecommunications and the Information Society, for its assistance in providing information, and co-ordinating and facilitating interviews with the relevant stakeholders.

Table of contents

Executive summary 7

Executive summary in Spanish 13

Introduction 21
    Information society strategies: Plan Avanza and Plan Avanza 2 22
    Scope and focus of this study 24
    Structure of the report 25
    Notes 26
    Bibliography 27

Chapter 1 Communication infrastructure 29
    Spain's telecommunication infrastructure in context 30
    Recent developments in the Spanish telecommunications market 31
    Assessment of Spanish action in communication infrastructure 35
    Universal service reform and the need for a national broadband plan 41
    Key assessments and proposals for action 59
    Notes 68
    Bibliography 70

Chapter 2 E-government: Reforming through information and communication technologies 73
    E-government – towards a paperless public administration 74
    The Spanish e-government strategy 77
    Digitising taxation, the early mover 85
    Digital justice or modernising for a new social deal 92
    Examples of good practice in e-government 100
    Key assessments and proposals for action 109
    Notes 118
    Bibliography 120

Tables

Table 1.1.	The New Avanza Infrastructure Programme	46
Table 1.2.	Spectrum tenders – final assignments	52
Table 2.1.	Budget dedicated to Plan Avanza 2, 2011	76
Table 2.2.	National ICT expenditures	80
Table 2.3.	Key transactions handled by the Tax Agency	88
Table 2.4.	Trends in IT costs relative to aggregate administrative costs	91

Figures

Figure 1.1.	Monthly LLU charges in the big EU member countries, end-2010	33
Figure 1.2.	Percentage of FTTH connections in total broadband in OECD countries, June 2011	37
Figure 1.3.	LLU access compared to bitstream and resale, 2005-2010	38
Figure 1.4.	Broadband market shares, 2005-2010	39
Figure 1.5.	Average monthly subscription (including line charge) for speeds between 2.5 and 15 mbps, September 2010 (USD PPP)	42
Figure 1.6.	OECD fixed (wired) broadband subscriptions by technology per 100 inhabitants, June 2011	42
Figure 1.7.	Broadband rural/urban availability	43
Figure 1.8.	DSL coverage and population density, 2009	44
Figure 2.1.	Framework of institutional co-ordination	79
Figure 2.2.	Implementing online service provision under Law 11/2007	81
Figure 2.3.	Percentage of citizens using the Internet to interact with public authorities, 2005 and 2010	83
Figure 2.4.	Percentage of businesses using the Internet to interact with public authorities, 2005 and 2010	84
Figure 2.5.	Online income returns in 2010	89
Figure 2.6.	Social media marketing: taxation through YouTube	90
Figure 2.7.	The development of expenditure in the Ministry of Justice	94
Figure 2.8.	Overall prioritisation of resources in the modernisation programme, EUR millions	95
Figure 2.9.	Social media platforms and channels for communications	97
Figure 2.10.	The three components of the judicial modernisation programme	99
Figure 2.11.	Maturity measurement for e-government	103
Figure 2.12.	Citizens' access to government services, Mexico	106

Executive summary:
Getting value out of information and communication technologies

Spain, hit hard by the global and European financial and economic crises, faces serious challenges. Low growth, unemployment rates above 20%, a severe fiscal imbalance, and budget challenges are key issues faced by the recently elected government. The new administration has clearly stressed the need to undertake the measures necessary to address the challenge of restoring growth to the economy and balance to the public sector budgets.

The new government has, in particular, reiterated the importance of using modern information and communication technologies (ICTs) to sustain efficient and effective public administration at all levels of government, as this may help to foster the greater overall competitiveness of the national economy. ICTs are vital to competitiveness and key enablers of economic growth and productivity. They also have an important social impact.

One key policy area has been improving communication infrastructure, and particularly the extension of broadband availability as a real incentive for next generation networks (NGN). Many consider this future-proof communication infrastructure a key foundation for economic growth and productivity.

Future administrative reforms are expected to support the elimination of inefficiencies and duplications, and to sustain the overall aims of downsizing, restructuring, and reducing the operating costs of the public sector.

E-government provides an important part of the answer by enabling higher efficiency and effectiveness in government administration and service delivery. The use of ICTs facilitates reductions in government expenditures by building on already existing investment and increasing efficiency and effectiveness. Leveraging the strategic use of e-government in the current economic context is imperative if public sector innovation and agility are to be increased, administration simplified, and service delivery and productivity improved. These insights were echoed in discussions at the last session of the OECD Public Governance Committee in the autumn of 2011.

This study does not include a comprehensive review of Plan Avanza 2. At the request of the Spanish government, the analysis focuses on selected areas of the plan that are linked to two of its overall objectives: improving communication infrastructure and the advent of paperless administration through the government use of ICTs. The new Spanish government is elaborating the Digital Agenda for Spain, which will be aligned with the Digital Agenda for Europe. The report reviews Plan Avanza 2, Spain's information society strategy. The study builds on existing data and background material and on information acquired through interviews conducted in July 2011 with the main stakeholders in Spain's public administration, the industry and academia.

Some common key challenges emerge from this analysis of communication infrastructure and the government's use of ICTs. They are:

- **Aligning strategies at all levels of government** to ensure coherence and exploit synergies. On one hand, this involves specifying how all the objectives set out in the Plan should be matched by operational initiatives that ensure the overall objectives are met. On the other hand, it means aligning efforts across all levels of government – national, regional and local.

- **Ensuring that resources are well prioritised,** allocated only to the soundest, most profitable investments, and subsequently spent effectively. Profitable investment may be ensured through better management and regulation of private markets and suppliers and through a systematic focus on how to support digital governance by using better data and professional ICT management and implementation processes.

- **Increasing user uptake**, and thus shifting focus from the supply to the demand of ICTs in order to benefit from the returns on previous investments. The significant achievements of the Spanish government in the development of ICT infrastructure and e-government service supply could be used more robustly as a platform to harvest financial and qualitative benefits.

Improving communication infrastructure

The Plan emphasises extending the availability of fixed and mobile networks throughout the country. Public funding of broadband infrastructure and universal service obligations have been combined with spectrum policy measures and other instruments, such as the new provisions for in-building wiring infrastructure, designed to remove barriers to network deployment.

High levels of co-ordination between the various measures undertaken as part of the Plan are desirable if impact is to be maximized. One of the principal strengths of the Plan is its internal policy coherence and its coherence with policy and regulatory regimes and institutional actors.

The European Digital Agenda: a national broadband plan. Spain expanded its universal service obligations as of 1 January 2012 to include broadband access at 1 Mbps. This move positions it as one of the few countries to have achieved the ambitious aim of guaranteeing broadband access through universal service obligations. Spain has thus met target I of the European Digital Agenda ("basic broadband for all"). The sequenced inclusion of broadband service as a universal service obligation was well designed, as public funding was made available to operators to extend coverage prior to complying with their obligations.

Spain has publicly endorsed the European Digital Agenda targets. However, it has not yet specified how these targets will be met, at least not in detail. Wireless technologies will certainly play a role in meeting them, at least in rural and remote areas. Nonetheless, Spain should provide a clearer interpretation of these targets and spell out what they actually mean in terms of communication infrastructure. Other enabling measures, in particular with respect to backhaul infrastructure, are likely to be needed. It is not clear, for example, how Spain will meet Digital Agenda targets II and III.

Spain should, therefore, conduct a comprehensive evaluation and set clear objectives, drawing up a roadmap and defining a set of available instruments to ensure that it meets the second and third Digital Agenda targets. One approach might be to draw up a national broadband plan which should specify: *i)* the specific quality of service requirements for the targets; *ii)* which technologies may meet the requirements for access, backhaul, and backbone; *iii)* whether the targets are to be met by market players only or if public funding will be necessary and/or available. Any public funding of the broadband network to meet the objectives should be clearly underpinned by a sound cost-benefit analysis.

Spain should maximize the benefits derived from the financial instruments recently put forward by the European Commission as part of its proposed Connecting Europe Facility funding arrangement. Furthermore, the current use of financial instruments for broadband infrastructure deployment under the Plan is thought to provide a reasonable leverage for the resources committed and shares common elements with the proposed Connecting Europe Facility. It is considered that these instruments minimise the market distortions that result from public intervention.

Broadband funding and market dynamics. Co-ordination between supply-side measures to improve communication infrastructure has been relatively successful. The supply-side input that had to be co-ordinated came from bodies such as the State Secretariat for Telecommunications and the Information Society (SETSI), Spain's Telecommunications Market Commission (CMT), the National Competition Commission (CNC) and the European Commission. The co-ordination of demand-side policies could, however, be improved. One possible area for improvement is the geographical alignment of broadband infrastructure funding and demand-side measures such as e-literacy or government use of ICTs. Spain should undertake a thorough assessment of the economic and social costs and benefits accrued through publicly funded broadband networks – for example, new business created or learning applications and healthcare services provided or facilitated.

Regulatory framework and market dynamics. A competitive broadband market and affordable prices are crucial to ensuring broadband take-up. Spain lags behind its OECD and European counterparts in both areas, even though the situation is improving. Recent developments in competition have also been partly based on off-price list discounting, which is not considered beneficial for consumers due to the lack of price transparency.

The Spanish regulator should, therefore, continue to encourage price-based competition in the broadband market. It should also make competition dynamics more transparent for consumers concerning off-price-list discounting. Moreover, Spain should transpose the new European regulatory framework into national law and the regulator should also undertake another round of communication market reviews. Both initiatives should be undertaken as rapidly as possible as they will help to give legal certainty to all stakeholders.

Spectrum reforms. Spain undertook substantial spectrum reforms for some years up to 2011. The measures were ambitious and forward-looking, and aimed at releasing the digital dividend band by 2015. The overall outcome so far has been well-designed and balanced in terms of the final spectrum holdings of all operators, public revenue, and market structure. Further involvement of smaller players would have been desirable. The authorities did in fact facilitate it, but without achieving the hoped-for participation. Coverage obligations have been introduced during the tender procedures, which will help to bring connectivity to rural areas. However, significant work still lies ahead as regards the release of the digital dividend by 2015.

Removing barriers to infrastructure deployment. The new in-building wiring regime for new buildings is considered excellent and will help remove barriers to infrastructure deployment for apartment buildings. Nonetheless, effects are likely to be visible in the long term only, given the low activity of the construction sector in Spain. Other measures could increase its effectiveness. They include extending or highlighting existing or future tax relief arrangements to cover fibre-ready buildings and reinforcing labelling schemes for buildings. Spain should also consider incorporating in-building wiring-related requirements into the general building review that takes place every 15 years.

Other positive measures for removing barriers to infrastructure deployment have been put in place. Among them are the symmetrical obligation for operators to allow access to in-building fibre infrastructure at reasonable prices and under transparent conditions, and the proposed requirement that they make ducts, poles, masts and land space available for the deployment of fibre and mobile communication infrastructure in newly-built railway and road infrastructure.

Focus on the value of e-government – consolidating and increasing user uptake

Policy reforms, as well as administrative reform efforts, can be supported and levered by good use of ICTs at all levels of government. Government use of ICTs and e-government are key enablers of successful modernisation processes. The following paragraphs advance observations and suggestions as to ways of improving the outcomes of e-government initiatives in Spain.

Strengthening digital governance to align strategic goals at all levels of government. E-government offers strong potential for addressing the challenges currently faced by the Spanish economy and society. In the past few years Spain has focused on implementing Law 11/2007 on citizens' rights to electronically access services. It seems largely to have accomplished its aim. On this basis, it could reflect and address the economic challenges it faces more forcibly in its current e-government orientations. ICT use can be leveraged to support and increase the impacts of public sector simplification and modernisation in the area of justice, for example, as stated by the new Prime Minister. Government policies could, and should, build on the use of ICTs to enhance and increase policy impact when relevant. However, the implementation of national policies and the use of ICTs still vary considerably across the autonomous communities and local authorities. To further its efforts, Spain is encouraged to consider:

- **Aligning e-government policies with public sector reform goals**, particularly to address budgetary and fiscal goals and challenges and to use ICTs to increase trust in government.

- **Strengthening the governance framework and developing a specific e-government strategy** beyond the Plan to support shared goal setting and co-ordination across all levels of government.

- **Comprehensively grounding the next e-government strategy** through international comparison with peers to obtain advice on good practice – in how to use ICTs to improve the business climate, for example.

Simplifying service supply and prioritising service channels. Spain has made significant progress in its online provision of public services and is now at the forefront of OECD countries in this regard. The approach of stating a set of citizens' rights with which public administration bodies must comply seems to have had a strong mobilising effect. However, the multi-channel service delivery strategy enabling citizens to choose their preferred communication and access channels seems to have fallen short of its aims. It fails to provide sufficiently clear incentives to support cost efficient prioritisation between different service delivery channels. Additionally, a digitised public administration is not by default a simpler, more trustworthy one: a more thorough understanding and greater participation in the use of public services – by businesses and citizens and stakeholders from all levels of government – could be instrumental in this regard in ensuring context-relevant and customised interaction. To further improve its service supply, Spain is encouraged to consider:

- **Better data to ground channel priorities.** Promote the use of indicators and data on service delivery channels to support prioritisation. Examples of data include those relating to cost structures, user preferences, and demand for services.

- **Stronger prioritisation of online channels.** Strengthen the use of the most efficient service delivery channels by, for example, increasing the mandatory use and/or provision of incentives for certain user groups.

- **Simplification and focus on user value.** Fundamentally challenge the traditional culture within administrative, regulatory and working processes and involve citizens and businesses.

Realising the value of ICTs by ensuring a higher e-government uptake. Spain seems to be aware of the key challenge that to fully benefit from its high online service provision it is important to increase user uptake of e-government services. Good practices like those of the Spanish Tax Agency might offer sound inspiration in this regard. The implementation model currently applied by the Ministry of Justice in the ongoing modernisation programme also seems to make good use of transparency as a benchmarking and implementation tool at local levels of government, particularly within the autonomous communities. Other models for engaging citizens to increase their uptake and trust could also be explored. In this regard, Spain is encouraged to consider:

- **Adopting common take-up measurements to ground policies.** One way might be the use of common models and data for assessing the delivery and uptake of services and supporting work on user uptake.

- **Marketing to increase awareness,** which involves strengthening and targeting the marketing and publicising of e-government services to achieve the highest impact – for example, by building on existing social networks.

- **Exploiting and enhancing ICT competencies, capacity building and training strategies.** One approach might be to build on existing cultural, educational and social policies targeting both citizens and businesses.

Consolidating government ICTs and adapting a whole-of-government approach. Spain cannot afford redundant infrastructure or services, missed synergies, or under-exploited data. Nor can it afford to invest in ICT projects of little value or in those whose benefits are not reaped. Clear management instruments identifying the added value in the use of ICTs are necessary. Considering public administration from a whole-of-government perspective enables seamless provision of services across levels of competency, shared services, and economies of scale. Spain has outlined a common government service architecture and seems to provide most essential shared services. It is a challenge – as it is across most OECD member countries – to ensure the full, co-ordinated uptake of these shared services at all levels of government, particularly in light of differences in ICT maturity and at both development and implementation levels. In order to address these challenges, Spain is encouraged to consider:

- **Standardising infrastructure and common components** to clarify further the government service and data architecture and define responsibilities across all levels of government.

- **Using business cases to focus on benefits realisation,** i.e. focus on the financial benefits of e-government and use business cases to determine the costs and benefits of key projects.

- **Consolidating and pursuing economies of scale,** as consolidating ICT infrastructure services can secure overall coherence and the realisation of potential economies of scale.

Executive summary in Spanish

Resumen ejecutivo: la optimización del uso de las TIC

Golpeada fuertemente por la crisis económica y financiera europea, así como por la crisis global, España se enfrenta a graves desafíos. Un crecimiento bajo, una tasa de desempleo por encima del 20%, un desequilibrio fiscal profundo y dificultades presupuestarias son cuestiones clave que debe confrontar el gobierno recién elegido. El nuevo ejecutivo ha insistido sin ambages en la necesidad de tomar las medidas necesarias para asumir el reto de recuperar el crecimiento de la economía y equilibrar el presupuesto del sector público.

El nuevo gobierno ha reiterado, en concreto, la importancia de utilizar las tecnologías de la información y la comunicación (TIC) modernas para respaldar una administración pública eficaz y efectiva en todos los niveles de gobierno, ya que así se podría fomentar una mayor competitividad del conjunto de la economía nacional. Las TIC son fundamentales para la competitividad y son impulsores clave del crecimiento económico y de la productividad. Asimismo, ocasionan un impacto social importante.

La mejora de la infraestructura de comunicación ha sido una política clave, en concreto la ampliación de la disponibilidad de la banda ancha a modo de incentivo real para las redes de nueva generación (RNG). Muchas personas consideran que esta infraestructura de comunicación de futuro es un fundamento esencial para el crecimiento económico y la productividad.

Se espera que las futuras reformas administrativas apoyen la eliminación de la ineficacia y las duplicaciones y respalden el propósito general de recortar, reestructurar y reducir los costes de funcionamiento del sector público.

La administración electrónica supone una parte importante de la respuesta ya que favorece una mayor eficiencia y eficacia de la administración pública y la prestación de servicios. Así pues, la utilización de las TIC facilita la reducción del gasto público al basarse en inversiones que ya existen e incrementar, de ese modo, la eficacia y la eficiencia. Es imperativo expandir el uso estratégico de la administración electrónica en el actual contexto económico si queremos multiplicar la innovación y la agilidad del sector público, simplificar la administración y mejorar la prestación de servicios y la productividad. Estas indicaciones se manifestaron durante los debates de la última reunión del Comité de Gobernanza Pública de la OCDE, celebrada en otoño de 2011.

Este estudio no incluye un examen exhaustivo del Plan Avanza 2. A petición del Gobierno español, el análisis se centra en aspectos seleccionados del plan que se vinculan con dos de sus objetivos generales: la mejora de la infraestructura de comunicación y el logro de una administración sin papeles mediante el uso de las TIC por parte del gobierno. El nuevo Gobierno español está en proceso de elaboración de la Agenda Digital para España que se adaptará a la Agenda Digital para Europa. En el informe se examina el Plan Avanza 2, la estrategia española de la sociedad de la información. El estudio se apoya en datos existentes y en documentación de contexto así como en la información obtenida durante las entrevistas que se realizaron en julio de 2011 con las principales partes interesadas de la administración pública española, de la industria y del mundo académico.

A partir del análisis de las infraestructuras de comunicación y el uso gubernamental de las TIC, surgen los siguientes retos clave comunes:

- **Armonización de estrategias en todos los niveles de gobierno** a fin de garantizar la coherencia y aprovechar las sinergias. Por una parte, esto implica especificar cómo se deberían vincular los objetivos diseñados en el Plan con las iniciativas de funcionamiento para garantizar que se cumplen los objetivos generales. Por otra, significa que es preciso que los esfuerzos discurran en paralelo en todas las esferas de la administración: nacional, regional y local.

- **Garantía de que los recursos están bien priorizados** y de que se asignan solo a las inversiones más rentables y convenientes y, con ello, se gastan de forma eficaz. La inversión rentable puede garantizarse a través de una mejor gestión y regulación de los mercados y proveedores privados, así como centrándose sistemáticamente en cómo apoyar la gobernanza digital mediante el uso de datos mejorados y procesos de gestión e implementación profesionales de las TIC.

- **Aumento de la captación de usuarios**, cambiando así la atención de la oferta a la demanda de las TIC a fin de beneficiarse del rendimiento de inversiones anteriores. Los significativos logros de España en cuanto al desarrollo de la infraestructura de las TIC y de la prestación del servicio de la administración electrónica podrían aprovecharse aún más como una plataforma para cosechar beneficios financieros y cualitativos.

Mejora de la infraestructura de comunicación

El Plan subraya la ampliación de la disponibilidad de redes fijas y móviles por todo el país. La inversión pública en infraestructura de banda ancha y la obligación de prestar un servicio universal se han combinado con medidas en materia de espectro radioeléctrico y con otros instrumentos, como las nuevas disposiciones sobre infraestructura de cableado de edificios, diseñadas para eliminar barreras al despliegue de la red.

Es deseable un mayor nivel de coordinación entre las distintas medidas tomadas en el marco del plan a fin de maximizar el impacto. Uno de los principales puntos fuertes del Plan es la coherencia interna de sus políticas, así como la cohesión con los regímenes reglamentarios y de políticas y los actores institucionales.

La Agenda Digital para Europa: un Plan Nacional de Banda Ancha. España amplió sus obligaciones de servicio universal a partir del 1 de enero de 2012 para incluir el acceso de banda ancha a una velocidad de 1 Mbps. Esto supone que es uno de los pocos países que ha logrado el ambicioso objetivo de garantizar el acceso de banda ancha mediante las obligaciones de servicio universal. Por lo tanto, España ha cumplido el objetivo I de la Agenda Digital para Europa («banda ancha básica para todos»). La inclusión secuencial del servicio de banda ancha como una obligación de servicio universal se ha diseñado adecuadamente ya que los operadores tuvieron acceso a financiación pública para ampliar la cobertura antes de cumplir con sus obligaciones.

España ha apoyado públicamente los objetivos de la Agenda Digital para Europa. Sin embargo, no ha especificado aún cómo se cumplirán dichos objetivos, o al menos no lo ha hecho de manera pormenorizada. Las tecnologías inalámbricas desempeñarán, sin duda, un papel en el logro de los objetivos, cuando menos en las zonas rurales y remotas. Con todo, España debería ofrecer una interpretación más clara de dichos objetivos y explicar las implicaciones con respecto a la infraestructura de comunicaciones. También es probable que sean necesarias otras medidas impulsoras, sobre todo por lo que se refiere a infraestructura de redes de backhaul. Por ejemplo, no queda claro qué va a hacer España para cumplir los objetivos II y III de la Agenda Digital.

Por consiguiente, España debería realizar una evaluación completa y establecer objetivos claros, además de redactar un plan de trabajo y definir un conjunto de instrumentos disponibles para garantizar que cumple con el segundo y tercer objetivo de la Agenda Digital. Una posibilidad sería elaborar un Plan Nacional de Banda Ancha, que debería especificar: a) los requisitos concretos de calidad de servicio de acuerdo con los objetivos; b) qué tecnologías podrían cumplir los requisitos en los niveles de acceso, backhaul y troncal; c) si los participantes del mercado por sí solos cumplirán los objetivos o si será necesaria, o estará disponible, la financiación pública. Cualquier financiación pública de la red de banda ancha a fin de cumplir con los objetivos debería estar claramente justificada por un sólido análisis coste-beneficio.

España debería aprovechar al máximo las ventajas derivadas de los instrumentos financieros presentados recientemente por la Comisión Europea en el marco del mecanismo de financiación propuesto «Conectar Europa». Asimismo, el uso actual de los instrumentos financieros para el despliegue de la infraestructura de banda ancha, conforme al Plan está pensado para proporcionar un aprovechamiento razonable de los recursos comprometidos y algunos de sus elementos son comunes al mecanismo "Conectar Europa" propuesto. Se considera que estos instrumentos minimizan las distorsiones del mercado resultantes de la intervención pública.

Financiación de la banda ancha y dinámica del mercado. La coordinación de las medidas relacionadas con la oferta para mejorar la infraestructura de comunicación ha sido relativamente satisfactoria. Las aportaciones del lado de la oferta que se tenían que coordinar procedieron de organismos como la Secretaría de Estado de Telecomunicaciones y para la Sociedad de la Información (SETSI), la Comisión del Mercado de las Telecomunicaciones de España (CMT), la Comisión Nacional de la Competencia (CNC) y la Comisión Europea. Sin embargo, podría mejorarse la coordinación de las políticas relacionadas con la demanda. Un posible ámbito de mejora sería la armonización geográfica de la financiación para la infraestructura de banda ancha y de las medidas relacionadas con la demanda, tales como la alfabetización electrónica y el uso de las TIC por parte del gobierno. España debería realizar una evaluación exhaustiva de los costes y beneficios sociales y económicos obtenidos a través de redes de banda ancha financiadas públicamente. Por ejemplo, las empresas que se han creado o las aplicaciones de aprendizaje y los servicios de asistencia sanitaria que se han proporcionado o facilitado.

Marco regulador y dinámica del mercado. Un mercado de banda ancho competitivo y unos precios asequibles son fundamentales para garantizar la aceptación de la banda ancha. España va con retraso en ambos aspectos con respecto a otros países europeos y de la OCDE, si bien la situación está mejorando. Los recientes avances en cuestión de competencia también se han basado en parte en descuentos de precios por debajo del precio de cabecera, que no se considera beneficioso para los consumidores debido a la falta de transparencia de los precios.

Por lo tanto, el regulador de España debería continuar fomentando la competencia basada en los precios para el mercado de la banda ancha. También debería hacer que la dinámica de la competencia fuese más transparente para los consumidores en cuestiones de descuentos de precios por debajo del precio de cabecera. Asimismo, España debería transponer el nuevo marco regulatorio europeo en su legislación nacional y el regulador también debería llevar a cabo otra ronda de análisis de los mercados de las comunicaciones. Ambas iniciativas deberían realizarse tan pronto como sea posible ya que servirán para conferir seguridad jurídica a todas las partes interesadas.

Reformas del espectro. España acometió sustanciales reformas del espectro durante varios años hasta 2011. Las medidas fueron ambiciosas y con miras de futuro y estaban orientadas a liberar la banda del dividendo digital para 2015. Hasta la fecha, el resultado global se ha diseñado bien y ha sido equilibrado en términos de explotación del espectro en su conjunto por parte de todos los operadores, los ingresos públicos y la estructura del mercado. Habría sido deseable contar con una mayor participación de los actores más pequeños. De hecho, las autoridades facilitaron dicha implicación pero sin lograrla en los términos deseados. Las obligaciones de cobertura se introdujeron durante los procedimientos de licitación y ayudarán a llevar la conectividad a las zonas rurales. Sin embargo, aún queda mucho por hacer con respecto a la liberación del dividendo digital para 2015.

Eliminación de barreras al despliegue de la infraestructura. El nuevo plan de cableado de los edificios para nuevas construcciones se considera excelente y servirá para eliminar barreras al despliegue de la infraestructura en los edificios de viviendas. Con todo, se espera que los efectos sean visibles solo a largo plazo dada la baja actividad del sector de la construcción en España. Otras medidas podrían aumentar su eficacia, entre las que se incluyen la ampliación o intensificación de medidas existentes o futuras de deducciones fiscales para los edificios preparados para la fibra óptica, y el refuerzo de los programas de etiquetado de los edificios. España también debería considerar la incorporación de requisitos de cableado en edificios en la inspección general de edificios que tiene lugar cada 15 años.

Se han puesto en marcha otras medidas positivas para la eliminación de barreras al despliegue de la infraestructura. Entre ellas se incluyen: la obligación simétrica de los operadores de permitir el acceso a las infraestructuras de fibra óptica en los edificios a precios razonables y con condiciones transparentes, y las obligaciones previstas para que haya conductos, torres, mástiles y espacio disponibles para el despliegue de la infraestructura de fibra óptica y comunicación móvil en infraestructuras ferroviarias y carreteras recién construidas.

El valor de la administración electrónica como punto central: consolidación e incremento del número de usuarios

Tanto las reformas de políticas como administrativas pueden ser respaldadas e impulsadas por un buen uso de las TIC en todos los niveles de gobierno. La utilización gubernamental de las TIC y la administración electrónica son impulsores clave de procesos de modernización fructíferos. Los siguientes párrafos presentan observaciones y sugerencias para mejorar los resultados de las iniciativas en el ámbito de la administración electrónica en España.

Fortalecimiento de la gobernanza digital para armonizar los objetivos estratégicos en todos los niveles de gobierno. La administración electrónica presenta un sólido potencial para abordar los retos a los que se enfrenta actualmente la economía y la sociedad españolas. En los últimos años, España se ha centrado en implementar la Ley 11/2007 de acceso electrónico de los ciudadanos a los Servicios Públicos. Parece que, en gran medida, ha logrado su objetivo. A partir de ahí, podría reflexionar y afrontar los retos económicos a los que se enfrenta de forma más contundente en sus actuales orientaciones sobre administración electrónica. El uso de las TIC puede armonizarse a fin de apoyar e incrementar el impacto de la simplificación y modernización del sector público en el ámbito de la justicia, por ejemplo, tal y como ha declarado el nuevo presidente del Gobierno. Las políticas gubernamentales podrían y deberían sustentarse en el uso de las TIC para mejorar e incrementar la repercusión de las políticas, siempre que resulte pertinente. Sin embargo, la ejecución de políticas nacionales y el uso de las TIC siguen variando considerablemente entre las Comunidades Autónomas y las autoridades locales. Para ampliar sus esfuerzos, se anima a España a considerar lo siguiente:

- Armonización de las políticas de administración electrónica con los objetivos de la reforma del sector público, especialmente a fin de acometer los objetivos y retos fiscales y presupuestarios y utilizar las TIC para que aumente la confianza en el gobierno.

- Fortalecimiento del marco de la gobernanza y desarrollo de una estrategia específica de administración electrónica más allá del Plan a fin de sustentar la definición de objetivos compartidos y la coordinación entre todos los niveles de gobierno.

- Justificación exhaustiva de la nueva estrategia de administración electrónica mediante la comparación internacional con otros países homólogos para obtener consejos sobre buenas prácticas; por ejemplo, cómo utilizar las TIC para mejorar el clima empresarial.

Simplificación de la prestación de servicios y establecimiento de prioridades con respecto a los canales de servicios. España ha logrado un avance significativo en su dotación en línea de servicios públicos y se encuentra ahora a la vanguardia de los países de la OCDE en este sentido. El hecho de haber presentado un conjunto de derechos de los ciudadanos que deben cumplir los organismos de la administración pública parece haber provocado un gran efecto movilizador. Sin embargo, la estrategia de la prestación de servicios a través de varios canales, que permite al ciudadano escoger sus canales de acceso y comunicación preferidos, no parece proporcionar incentivos lo suficientemente claros como para apoyar la definición de la rentabilidad a fin de establecer prioridades entre los distintos canales de prestación de servicios. Al mismo tiempo, una administración pública digitalizada no supone, por sí sola, que los servicios públicos sean más sencillos ni que generen más confianza. Es decir, una mayor comprensión y una participación más amplia en el uso de los servicios públicos (por parte de empresas y ciudadanos, así como de los actores interesados en todos los niveles de gobierno) podrían ser instrumentales en este sentido para garantizar una interacción personalizada y adaptada al contexto. Para seguir mejorando la prestación de servicios, se anima a España a considerar lo siguiente:

- Datos mejorados para justificar las priorización de los canales. Promoción del uso de indicadores y datos sobre canales de prestación de servicios para apoyar el establecimiento de prioridades. Por ejemplo, los datos incluyen información relativa a las estructuras de costes, preferencias de los usuarios y demanda de servicios.

- Una mayor priorización de los canales en línea. Refuerzo de la utilización de los canales de prestación de servicios más eficaces, por ejemplo, mediante el incremento del uso obligatorio o la provisión de incentivos para ciertos grupos de usuarios.

- Simplificación y concentración en el valor del usuario. Puesta a prueba fundamentalmente de la cultura tradicional que impera en los procesos administrativos, reguladores y de trabajo e implicación de los ciudadanos y de las empresas.

Aprovechamiento del valor de las TIC garantizando, para ello, un mayor uso de la administración electrónica.

España parece ser consciente del desafío clave que supone que para aprovechar al máximo la prestación de servicios en línea es importante aumentar el número de usuarios que utilizan los servicios de la administración electrónica. Las buenas prácticas, como las aplicadas por la Agencia Tributaria española, podrían servir de inspiración a este respecto. El modelo de implementación que aplica actualmente el Ministerio de Justicia en el programa de modernización en curso también parece estar aprovechando la transparencia como una herramienta de referencia y de aplicación en los niveles locales de gobierno, sobre todo en las Comunidades Autónomas. También se podrían estudiar otros modelos para aumentar la participación y la confianza de los ciudadanos. En este sentido, se anima a España a considerar lo siguiente:

- Adopción de medidas comunes de captación de usuarios para justificar las políticas. Una podría ser el uso de modelos y datos en común para evaluar la prestación y la utilización de servicios y apoyar las labores de captación de usuarios.

- Marketing para aumentar la conciencia, que implica fortalecer y orientar el marketing y la publicidad a los servicios de la administración electrónica para alcanzar la mayor repercusión posible; por ejemplo, aprovechar las redes sociales existentes.

- Explotación y mejora de las competencias en materia de TIC, capacitación y estrategias de formación. Una de las medidas podría ser aprovechar las políticas existentes en el ámbito cultural, educativo y social destinadas tanto a los ciudadanos como a las empresas.

Consolidación de las TIC gubernamentales y adopción de una perspectiva pública común. España no se puede permitir la duplicación de infraestructuras, la prestación de servicios, las sinergias no aprovechadas o el actual aprovechamiento deficiente de los datos. Tampoco se puede permitir la inversión en proyectos de TIC de valor reducido o aquellos de los que no se obtengan beneficios. Son necesarios instrumentos de gestión claros que identifiquen el valor añadido en el uso de las TIC. Considerar la administración pública desde una perspectiva común que abarque al conjunto del gobierno hace posible la prestación de servicios carente de problemas en todos los niveles competenciales, los servicios compartidos y las economías de escala. España ha esbozado una estructura común de servicios gubernamentales y parece ofrecer la mayoría de los servicios compartidos esenciales. Al igual que en la mayoría de los países miembros de la OCDE, supone un reto garantizar la utilización plena y coordinada de estos servicios compartidos en todos los niveles de gobierno, especialmente a la vista de las diferencias en cuanto a la madurez de las TIC tanto a escala de desarrollo como de implantación. Para abordar estos desafíos, se anima a España a considerar lo siguiente:

- Estandarización de infraestructuras y componentes comunes a fin de seguir clarificando los servicios prestados por el gobierno y la arquitectura de datos, y de definir las responsabilidades en todos los niveles de gobierno.

- Uso de ejemplos empresariales para centrarse en la obtención de beneficios, es decir, centrarse en las ventajas financieras de la administración electrónica y utilizar ejemplos del ámbito empresarial para determinar los costes y las beneficios de los proyectos clave.

- Consolidación y búsqueda de economías de escala, ya que la consolidación de los servicios prestados por la infraestructura de las TIC puede garantizar la coherencia generalizada y la consecución de economías de escala en potencia.

Introduction

Information and communication technologies, widely known as ICTs, have radically changed our societies. No economy can be competitive today without efficient supply and use of ICTs. Nor can policy makers hope to foster social and economic development unless they fully embrace it.

Several terms are used to designate the growing use of ICTs. One such term, the "Internet economy", covers the full range of economic, social, and cultural activities supported by the Internet and related technologies. It was in recognition of the need to further the development of the Internet economy that ministers and representatives of OECD countries, meeting in the Korean capital, Seoul, adopted the Seoul Declaration for the Future of the Internet Economy on 18 June 2008. The declaration, which forms the basis for action by OECD countries to strengthen the Internet economy, states its intent as follows:

> We **share** a vision that the Internet economy ... will strengthen our capacity to improve the quality of life for all our citizens ... by providing new opportunities for employment, productivity, education, health and public services as well as addressing environmental and demographic concerns.

"Information society" is another term that has been coined to describe how ICTs have come to shape society. It commonly refers to societies where ICTs play a central role, affecting every walk of life, and where creating, distributing, and manipulating information have become significant activities. Governments of information societies face the daunting challenge of harnessing ICT to the effective, efficient delivery of public policies and services. Spain is one such country.

In 2008, after a decade of rapid growth, the country entered recession, like other European economies (OECD, 2010c). The result in Spain has been a massive rise in unemployment to a level that has now passed 20%. Severe fiscal imbalances and budget challenges, an increasing gross debt and low growth are among key issues. Instability and rising interest rates in the euro area reinforce the need to ensure trust in Spanish economic policies and government administration (OECD, 2011a). Spain has already undertaken far-reaching measures, reducing expenditure, public sector wages and staffing, and public infrastructure investment (OECD, 2011b; 2011a). Policy responses are urgently required both to consolidate public finances and restore economic growth.

Information society policies and ICT are considered important to Spanish economic recovery and growth (OECD, 2010c). Indeed, Spanish policy makers and stakeholders also believe that information society polices and ICT may be an instrument of societal change and the advent of a competitive, equitable, sustainable knowledge-based economy. However, crucial to any successful ICT-related initiative are regional and local implementation and cross-regional co-ordination. Spain has a decentralised multi-level administrative structure with regional governments that boast a broad span of competencies and account for roughly one-third of all public expenditure

(OECD, 2011b). Some 8 000 municipalities and other entities form the local level of government.

Information society strategies: Plan Avanza and Plan Avanza 2

Plan Avanza

Plan Avanza, the previous information society strategy, was Spain's umbrella strategy for the advancement of an information society. Launched in 2005 to cover the period 2006-2010, it addressed action in four areas: digital citizenship, digital economy, digital public services, and digital context. It provided a major, concerted policy response to the role to be played by ICT in the Spanish economy and society. Some of its milestones were public sector modernisation, increased broadband availability, ICT diffusion, and the emergence of the ICT sector as a driver of growth.

In 2010 the OECD reviewed the previous information society strategy (OECD, 2010b) and identified three areas of action: leadership and commitment as key drivers; progress in ICT diffusion and access; and a move towards greater value creation from ICTs. It framed its proposals for future policy within these three themes (OECD, 2010c), recommending that the previous plan should:

- strengthen the contribution of ICTs to economic recovery and long-term objectives for green growth and innovation;

- support a demand-driven, user-centred strategic approach to public service delivery;

- maximise the impact of initiatives while making optimal use of resources invested.

These policy recommendations have contributed to Spain's continuous adjustment of its efforts to put in place an information society strategy and to its finalisation of Plan Avanza 2.

The OECD described the previous information society strategy as the most ambitious and comprehensive policy that the government of Spain had undertaken towards the advancement of an information society (OECD, 2010b). The OECD also asserts in its initial review that the plan has an important role to play in the current economic context. In the short term, it could make an especially important contribution to Spain's economic recovery and, in the long term, play a crucial part in promoting sustainable economic growth.

Plan Avanza 2

On 16 July 2010, the Spanish Council of Ministers approved the second stage of Plan Avanza,: the 2011-2015 Strategy for Plan Avanza 2. It is designed to ensure continuity with the original plan's course of action. It includes projects in progress and updates initial objectives so that they respond to the new challenges of the information society.

One of the Plan's prime objectives is to leverage information and communication technologies to change Spain's economic model. The thinking is that wider use of ICTs can help increase competitiveness and productivity and favour equal opportunities across the country, so boosting the economy and consolidating a sustainable model of economic growth.

The previous information society strategy, aimed at catching up with European Union (EU) averages, especially in respect of coverage and connectivity. The second stage seeks to make Spain a leader in the development and use of advanced ICT products and services.

The Plan's structure and strategy implementation plan for 2011-2015 have been approved. The State Secretariat for Telecommunications and the Information Society (SETSI) under the aegis of the Ministry of Industry, Tourism and Trade co-ordinates the implementation of information society strategy centrally. It seeks co-operation with the private sector and other social, political, and institutional stakeholders in order to ensure efficient, effective implementation. Implementation is organised around a collaborative framework that brings together all levels of Spanish government (central, regional, and local), industry, and other representatives of society. Spain's Telecommunications and Information Society Advisory Board (CATSI), for example, provides input on prioritising the use of the Plan's resources.[1] Because this information society strategy involves Spanish society as a whole, the government regards its development as a priority, particularly as it aligns with a number of initiatives currently being undertaken in the EU.

On 19 May 2010, the European Commission adopted a Communication on the Digital Agenda for Europe, one of the seven flagship initiatives of the Europe 2020 Strategy. The Agenda, which follows on from i2010, sets out the key enabling role ICTs will have to play if Europe is to succeed in its ambitions for 2020. Its objective is to maximise the contribution that the information society and ICT development can make to economic recovery and job creation in the European Union by 2015. The Plan's ambition translates into four strategic axes: Infrastructure, Trust and Security, Paperless Administration, and ICT sector development.[2]

Spain has identified 34 key challenges in the field of ICTs which it has to face. The Plan's strategy document defines how to address these challenges by aiming to achieve ten objectives outlined in Box 0.1.

Box 0.1. **The Plan's ten concrete objectives**

1. Promoting innovative ICT processes in the public administration.
2. Spreading ICTs in healthcare and for welfare.
3. Modernising the education and training model through the use of ICTs.
4. Spreading telecommunication networks and increasing their capacity.
5. Spreading trustworthy ICTs among citizens and enterprises.
6. Increasing the advanced use of ICT solutions among citizens.
7. Spreading the use of ICT business solutions in enterprises.
8. Developing technological skills in the ICT sector.
9. Strengthening the digital content sector and intellectual property rights in the current technological context and within the Spanish and European legal framework.
10. Developing green ICTs.

Source: Spanish Ministry of Industry, Tourism and Trade (2010), *Estrategia 2011-2015, Plan Avanza 2*, Ministry of Industry, Tourism and Trade, Madrid.

The government has identified over 100 measures that can be taken to achieve the 10 objectives, as well as performance indicators for measuring results. It has also drawn up a set of reforms scrapping rules and regulations that hinder the expansion and use of ICTs and guaranteeing the rights of citizens in an information society.

The 2011-2015 Strategy for Plan Avanza 2 is to maintain the collaborative model with the regions (*Comunidades Autónomas* and *Ciudades Autónomas*), local government, and public and private institutions and enterprises, used in the previous information society strategy.

Scope and focus of this study

The present strategic study of the Plan examines two important issues, selected to underline the need for successful supply and demand-side policies in building a far-reaching, inclusive, and efficient information society. They are:

- Improving communication infrastructure, consideration of which involves assessing some of the main measures taken to extend communication networks, particularly broadband networks.

- Paperless administration, considered as a contribution to achieving efficiency gains and savings within the Spanish public sector and an ubiquitous, enhanced provision of public services. It is addressed through the prism of the paperless administration of justice and taxation.

Improving communication infrastructure

The fourth of the Plan's 10 objectives (Box 0.1) relates to improving the capacity and coverage of telecommunication networks: ultra high-speed fixed and mobile networks should be deployed to support the development of the information society.

The ultimate goal of most OECD infrastructure policy makers, including Spain's, is the development of broadband networks and services to attain the greatest practical coverage and use. Ultra high-speed fixed broadband networks require heavy investment, especially those that are fibre-based. A high policy priority is therefore to create a legal framework that provides incentives and certainty for investment in fibre broadband networks.

Chapter 1 addresses the improvement of communication infrastructure. It assesses the policies that Spain has implemented to date and makes a set of recommendations designed to assist it in pursuing its policies. This assessment will be informed by a strategic analysis of the public policy approach to upgrading infrastructure in Spain, illustrated where necessary by selected best practices that OECD countries have undertaken.

Chapter 1 will also consider the constraints with which Spain has to contend. They include some of the goals it has assigned to its policies, such as:

- the creation of a legal and economic framework that provides incentives for investment in next generation access (NGA) networks;

- public funding for the deployment of NGA networks in rural, remote, and less privileged areas;

- the increased availability of spectrum resources for wireless broadband services;

- the removal of obsolete in-building wiring infrastructure which obstructs the rollout of NGA networks.

Paperless administration

The Spanish government considers ICTs an innovative catalyst for achieving a more efficient, sustainable, citizen-focused public sector, where ICTs support productivity growth and reductions in expenditures. One important challenge addressed by the Plan is to foster e-government as part of a bid to achieve a paperless public administration by 2015. A paperless administration entails computerising and automating administrative procedures; digitalising data, records, and services; ensuring greater uptake of digital public services by citizens and businesses; reducing existing ICT costs; and the building of a comprehensive digital infrastructure able to support all such measures.

In addition, government departments have their own goals. Specific modernisation goals have been set for the judicial system as part of the current Strategy Plan for Modernisation of the Spanish Justice System 2009-2012. The Tax Administration has established objectives to meet the strategic challenges of the Spanish Tax Agency and its users, e.g. to combat social security fraud and increase efficiency.

Chapter 2 addresses e-government and paperless administration through four themes:

1. the application of the strategy and co-ordination framework applied, looking for example at institutional structures and budgets;
2. service delivery channels and the supply of digital services with special emphasis on the maturity of services and their delivery;
3. the demand for online services and user uptake, focusing on the actual use of the services provided;
4. consolidation and realisation of the benefits of e-government,

Benchmarks with OECD and EU member countries are used throughout the analysis to support the arguments.

Structure of the report

The report is structured according to the two analytical perspectives outlined above. The OECD Directorate for Science, Technology and Industry has prepared the chapter on communication infrastructure (Chapter 1) and the Directorate for Public Governance and Territorial Development the one on e-government (Chapter 2).

Chapter 1 comprises four sections. They look at the background, consider the Spanish market context, then analyse and assess the communication infrastructure measures taken. A comprehensive conclusion sets out key findings and recommends proposals for action.

Chapter 2 is devoted to e-government and focuses particularly on the areas of e-justice and e-taxation. It comprises five sections, beginning with an overview of Spain's e-government strategy and co-ordination framework. It then outlines and puts into context the approaches to digitising the tax and the justice systems. The following section considers relevant good practices from OECD member countries that may support the Spanish government in tackling the key challenges it has identified. The chapter concludes with key findings and proposals for action.

Notes

1. CATSI, the *Consejo Asesor de Telecomunicaciones y Sociedad de la Información*, is established by law, *e.g.* the Royal Decree 1029/2002. See OECD (2010) for a detailed description.

2. The initial Plan Avanza 2 mentions 5 strategic axis. However the two axes, "Technological Training and Content and Services", have been merged with "Paperless Administration".

Bibliography

OECD (2010a), *Good Governance for Digital Policies: How to Get the Most Out of ICT: The Case of Spain's Plan Avanza*, OECD Information Society Reviews, OECD Publishing, Paris, *http://dx.doi.org/10.1787/9789264031104-en*.

OECD (2010b), *OECD Information Technology Outlook 2010*, OECD Publishing, Paris, *http://dx.doi.org/10.1787/it_outlook-2010-en*.

OECD (2010c), *OECD Economic Surveys: Spain 2010*, OECD Publishing, Paris, *http://dx.doi.org/10.1787/eco_surveys-esp-2010-en*.

OECD (2010d), "Survey Report: Survey of Trends and Developments in the Use of Electronic Services for Taxpayer Service Delivery, Forum on Tax Administration's Taxpayer Services Subgroup", OECD, Paris, *www.oecd.org/ctp/taxadministration/45035933.pdf*.

OECD (2011a), "OECD perspectives: Spain, policies for a sustainable recovery", *www.oecd.org/spain/44686629.pdf*.

OECD (2011b), *Restoring Public Finances*, Special Issue of the *OECD Journal on Budgeting*, Volume 2011/2, OECD Publishing, Paris, *http://dx.doi.org/10.1787/budget-v11-2-en*.

Spanish Ministry of Industry, Tourism and Trade (2010), *Estrategia 2011-2015, Plan Avanza 2*, Ministry of Industry, Tourism and Trade, Madrid.

Chapter 1

Communication infrastructure

This chapter examines the measures the Spanish government has taken to reform and extend Spain's broadband communication infrastructure under the Plan. It first sets them against the broader context of current developments in the market and examines related domestic and European policies from the overall perspective of improving Spain's network connectivity.

The chapter then goes on to explore and assess the different actions and technical options Spain has taken. It principally scrutinises the country's supply-side policies, but also discusses how they may be aligned with action on the demand side. It stresses the importance of a carefully co-ordinated, whole-of-government strategy and a technologically holistic approach that should constantly monitor progress (or lack of).

The chapter concludes with a review of the key assessments to emerge from the analysis and proposes recommended courses of action where necessary.

Spain's telecommunication infrastructure in context

This chapter examines the provisions in the Plan for deploying a communications[1] infrastructure. Like the 2006-10 stage, it aims to bring broadband networks to outlying municipalities, particularly in rural areas with low population density where there is an infrastructure and connectivity deficit.

One difficulty in implementing information society strategies is striking the right balance between supply- and demand-side approaches. Although they are complementary and both kinds are needed to promote the virtuous circle of ICT availability, adoption and use, it is natural that supply-side measures should be more prominent in the first stages of any information society strategy. Once the infrastructure is in place, however, there is a gradual shift to the demand side.

The Plan places a greater emphasis on the demand side than its predecessor, although it does acknowledge that an upgraded communication infrastructure remains a key enabler for other electronic services and applications. Indeed, its fourth objective (Box 1.1) is to upgrade the capacity and coverage of communications networks, especially of very high-speed ones (both fixed and mobile). In this respect, the Spanish government has taken a number of measures.

- It included basic broadband service with 1 Mbps download speed in universal service obligations as of 1 January 2012. It did the same for public funding of broadband infrastructure – both access and backhaul – for broadband extension and rollouts of next generation access (NGA) broadband infrastructure (at speeds higher than 50 Mbps).

- It auctioned a total of 310 MHz of spectrum, including digital dividend spectrum, in 2011 as part of a policy to sell off spectrum. It used refarming and comparative selection processes for the 900 MHz and 1 800 MHz band and a spectrum auction for the 800 MHz, 900 MHz, 1 800 MHz and 2.6 GHz bands. As a result, available spectrum resources for mobile communications have increased by 70%, EUR 2 billion was raised in public revenue, and the foundations for efficient long-term market structures were laid.

- It revised the legal framework for common in-building wiring infrastructure in multi-dwelling units, introducing a new regulation2 in 2011 to ensure that the infrastructure (i.e. fibre-to-the-home-and-building network architecture) for very high speed communication served multi-dwelling premises.

Against this background, the following sections go on to offer an insight into the Spanish telecommunications market before considering it in the broader European and OECD contexts. The chapter then considers ways to:

- create a legal and economic framework that offers incentives to invest in NGA broadband networks and maintain healthy levels of competition;

- make efficient use of public funding so as to maximise the coverage of broadband networks, particularly in rural, remote, and less-privileged areas;

- remove existing barriers to the deployment of NGA networks (e.g. those that hinder necessary in-building wiring infrastructure);

- reform spectrum policy in order to promote its efficient use, make sufficient resources available for wireless broadband services, and foster competition and innovation in the wireless industry.

The chapter concludes with an assessment of policy measures adopted so far, cites best practices where they illustrate relevant arrangements in place in other countries, and provides a set of recommendations for Spain's information society strategy. It chiefly considers supply-side policy measures, referring to those on the demand side only when they may be co-ordinated with network development measures.

Recent developments in the Spanish telecommunications market

In December 2010, the Spanish government proposed two important pieces of legislation: a General Telecommunications Bill,[3] which transposes the revised regulatory framework for electronic communications, and a draft royal decree[4] on the implementation of refarming in the 900 MHz and 1 800 MHz wavebands and the assignment of spectrum in the 800 MHz, 900 MHz, 1 800 MHz and 2.6 GHz bands. Although the new draft General Telecommunications Bill has not been passed yet due to the dissolution of Parliament and the subsequent legislative election of December 2011, two royal decrees authorising radio spectrum reforms were issued in 2011.

Electronic communications issues have also been addressed in other pieces of legislation. A provision of the Law on a Sustainable Economy, approved in March 2011,[5] defines 1 Mbps as the minimum rate of downstream traffic required for functional Internet access. This requirement was to be included under the universal service obligation from the beginning of 2012. The Law on a Sustainable Economy also touches on issues like institutional design, lower administrative charges and, most importantly, specifications governing the radio spectrum regime. Of particular significance are the law's provisions on the use of the digital dividend frequencies for wireless broadband delivery and bringing the main mobile service frequency bands within the scope of secondary trading.

Regarding communications regulation,[6] there is a division of regulatory functions between different national regulatory authorities (NRAs). Two bodies, however, share regulatory responsibility: the Telecommunications Market Commission (CMT) and the State Secretariat for Telecommunications and the Information Society (SETSI), which comes under the aegis of the Ministry of Industry, Energy and Tourism,[7] concurrently responsible for electronic communications policy. The National Competition Commission (CNC) deals with *ex post* competition policy (e.g. anti-trust and merger reviews), unless matters are on such a scale that they are referred to the European Commission. The CNC is also required to provide non-binding advice on policy measures that may affect competition in the national market (market reviews, legal proposals, etc.).

Network development in Spain

Telefónica, the incumbent operator, continues to dominate the Spanish fixed telephony retail market. Its market share as an incumbent operator is converging towards EU averages after dropping in terms of revenue (from 79% in December 2008 to 68.4% in December 2010) and traffic (from 67.5% in December 2008 to 58.5% in December 2010).

Mobile broadband penetration reached 42.4% in June 2011, just below the OECD average of 47.9%. The number of fixed broadband lines has grown, too, albeit at a slower pace than in previous years. Penetration, as measured by lines per 100 inhabitants, increased from 21.5% in January 2010 to 23.7% in June 2011 and 24.1% in November 2011. It nevertheless remains below the OECD average of 25.1%.

In Spain, 90.75% of broadband connections are faster than the advertised download speed of 2 Mbps, and one out of three connections is equal to or greater than 10 Mbps. With four out of five fixed broadband lines, DSL continues to be the main technology for the provision of fixed broadband services. Cable comes next. Market shares have remained more or less constant: as of November 2011, alternative DSL operators accounted for 31.08% of the whole broadband market (and 40% of the DSL market) and cable operators 19.52%. Throughout 2011, the incumbent's market share slipped significantly, falling to 49.40% in November 2011. The drop came after a three-year period when its market share had actually increased – from 53.8% in 2005 to 56.4% in 2008 (CMT, 2010a; CMT, 2011).

As a member country of the European Union, Spain abides by the European framework for electronic communications.[8] Under the framework, *ex ante* economic regulation is applied in order to promote competition in the interest of consumers while ensuring sufficient network investment. The main provisions are administered through economic regulation of market operators who have been declared to have "significant market power" (SMP). In the Spanish fixed-access market, only the incumbent Telefónica is considered to have SMP, while cable operators, including ONO, stay outside the defined market. Regulations grant alternative operators wholesale access to Telefónica's network, which promotes competition.

Broadband access options

In broad terms, there are two main wholesale broadband access products: local loop unbundling (LLU) and indirect, or bitstream, access. LLU may take the form of fully unbundled access, shared access, or LLU without basic telephone service (known as "naked DSL"). It gives entrant operators access to the copper loop of the incumbent operator, which allows them to create a competing access network by renting the incumbent's local loop at regulated prices. Notionally, therefore, the incumbent's economies of scale are available to all the operators in the market. Another access product, such as bitstream, can be used without alternative providers having to build a line out to every local exchange.

Local loop unbundling is the preferred option of alternative DSL operators, followed by bitstream access and (very few) resale lines. Between January 2010 and November 2011, the number of fully unbundled loops increased from around 1.7 million to over 2.6 million. The rise was, in part, due to the fact many shared access lines were migrated to the "naked DSL". Under this unbundling arrangement, the alternative provider pays the incumbent operator a fee equivalent to the fully unbundled fee to provide broadband and voice-over-Internet-protocol (VoIP) services after the subscriber has cancelled the incumbent's fixed telephony services (and no longer pays a monthly line rental).

The development of LLU as the preferred means of access coincides with Spain's deep cuts in wholesale charges, bringing them down to the European average and making them the lowest of those in the large European Union member countries (Figure 1.1).

Figure 1.1. **Monthly LLU charges in the big EU member countries, end-2010**

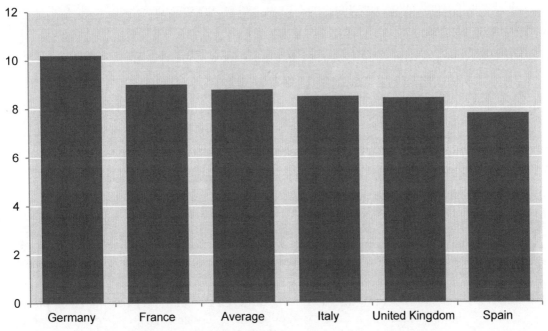

Source: Country notifications to the European Commission.

Wholesale broadband products currently offer speeds of up to 30 Mbps. In November 2010, the Spanish regulatory body, the CMT, defined a new region-wide wholesale broadband service that goes under the acronym NEBA, for new ethernet broadband access. It was the fruit of joint effort that brought together the CMT and the incumbent and alternative operators, who began work after approving an analysis of the wholesale broadband access market, known as "Market 5". NEBA is intended to replace existing asynchronous transfer mode (ATM) and IP bitstream products. The CMT board approved it on 16 November 2011[9] and operators should be taking it up it in the first quarter of 2012. Some alternative providers have complained about the delays in NEBA's deployment, as they perceive it as a wholesale service essential to the maintenance of competition in the market.

The reported level of retail broadband prices is high according to OECD, EU and CMT studies, although competition has grown since 2009 and prices have seemingly dropped. This chapter provides further analysis of price and competition dynamics in the Spanish broadband market below.

There have not yet been any major FTTH deployments in Spain, which lags behind some advanced OECD countries but is in line with such major countries as Australia, Canada, France, Germany, and the United Kingdom. In August 2010, the incumbent launched new FTTH broadband offers in a restricted number of areas. They boasted speeds of up to 50 Mbps, although most cable operators offer rates of 100 Mbps. While still relatively low, FTTH uptake is increasing in Spain. In November 2011, the number of FTTH subscribers stood at 158 204, a 247.9% year-on-year increase. ONO, the largest cable operator, has upgraded its network and has the capacity to provide 50% of the Spanish population with FTTC services in its coverage area. Three regional cable operators (R in Galicia, Telecable in the Asturias, and Euskaltel in the Basque Country) complete the picture with a national coverage of 60%.

Local authorities in Spain run several publicly financed networks, most of which use WiFi technology and some fibre optics. The CMT recently issued rules on the conditions under which local public entities may fund and/or operate broadband networks. As regards public aid for broadband networks in the form of loans, EUR 133 million were allocated in 2010 under the Plan, while in 2011 the financing programme freed up EUR 100 million, although only EUR 31 million were actually allocated.

Mobile telephony

The mobile penetration rate in Spain was 126.5% in September 2011. When compared with OECD countries in 2009, it ranked 19th of 34 countries (111.2 subscriptions in Spain compared to an OECD average of 102.6 subscribers).

As for mobile broadband service providers, their market shares measured in numbers of subscribers had remained relatively stable until 2007. In that year the mobile market was opened to virtual network operators (MVNOs). They, together with the fourth network operator Yoigo, which started operating in 2006, consistently gained market shares. By November 2011 Yoigo's share was 6.3% (or 5.13% measured in numbers of subscribers), while the market leader (the mobile telephony arm of the fixed incumbent) held a 40.09% share and its main competitor Vodafone 28.12%.

Large-scale deployment of long-term evolution (LTE) technology has not taken place yet, even though Telefónica and Vodafone did conduct trials or provide large corporations with pre-commercial services in the second half of 2011. Deployments on a larger scale are expected to take place in early 2012.

Rights of way and facility sharing

There have been two significant moves relating to facility sharing in Spain. The first, in accordance with requirements introduced by the CMT in 2008 (Telefónica, 2011), is that the incumbent operator should publish its reference offer for sharing ducts and poles. The offer should be at cost-oriented prices and contain non-discriminatory, transparent conditions. The second move was to place the same obligations on all operators for the sharing of in-building fibre infrastructure.[10] Thus, the first operator to reach a building with its fibre network, regardless of its SMP, should meet reasonable access requests by third parties at reasonable prices and under transparent conditions.

In addition, the Advisory Committee for the Deployment of Very High Speed Access Infrastructure (CADIAU) advised SETSI on the drafting of two pieces of legislation designed to facilitate network deployment. One is related to communal infrastructure for in-building telecom services and upgraded earlier regulations to the NGA context.[11] The second, parallel, piece of legislation pertained to the deployment of infrastructure serving public works on roads and railway lines.[12]

Universal service

As an EU member country, Spain commits under the Universal Services Directive on e-communications to provide the "minimum set of services of specified quality to which all end-users have access, at an affordable price in the light of specific national conditions, without distorting competition".

The main elements in calculating the net cost of universal e-communication needs continue to be fixed access and arrangements to meet special needs and low incomes. The four main operators lodged appeals against the net cost calculation and the CMT's decision of August 2008 to require them to bear the cost of the universal service compensation mechanism for the years 2003-2005. The appeals are still pending in national courts, although the operators paid up in April 2009.

Bids for the provision of universal service were tendered in 2011, with Telefónica securing the contract in November of the same year and being designated universal service provider until 31 December 2016. Under the terms of Law on a Sustainable Economy, its bid includes 1 Mbps of effective download speed, considered the minimum for ensuring functional Internet access as part of the universal service provision.

The European context

On 19 May 2010, the European Commission adopted a Communication on the Digital Agenda for Europe.[13] It is one of the seven flagship initiatives of the Europe 2020 growth strategy, launched in March 2010 with the aim of preparing Europe's economy for the coming decade.[14] The agenda states that ICTs can contribute to the strategic vision of Europe 2020 by creating advanced communication networks. The agenda sets out seven action areas, one of which is the importance of widening Europeans' access to fast and ultra-fast Internet. Its stated targets are to give all Europeans access to Internet speeds of 30 Mbps or higher and to ensure that at least 50% of European households subscribe to Internet connections of 100 Mbps and higher.

In broad terms, the European Commission believes it can attract investment in broadband through better, more consistent regulation and operational schemes such as credit enhancement mechanisms and guidance on attracting investment in fibre-based networks. The identified instrument for meeting these goals is the European Commission's Recommendation on Regulated Access to Next Generation Access Networks and its commitment to offering guidance on costing methodologies.

Spain endorsed Europe 2020 and its Digital Agenda for Europe in the Granada Ministerial Declaration[15] and, as a member of the Council of the European Parliament, also adopted the Digital Agenda in its own right (Council of the European Union, 2010).

Assessment of Spanish action in communication infrastructure

A point about the Plan that should be restated is that it is the second stage in an original plan and is part of a broader set of measures to further Spain's network development. The measures under consideration here pertain essentially to the European and national regulatory frameworks and the decisions taken by the CMT.

This section also examines various components of the Plan and some of the strengths and weaknesses the model pursues. Yet identifying "best practices" in broadband and, in particular, in NGA deployment is a challenge – not least because there is no set best practice in this area. Even where network migration to NGA has been largely completed or is in the process of being so, many uncertainties remain as to the questions like the efficiency of investment and technology choices. In that light, any assessment is at best partial.

However, it is entirely possible to examine whether the measures adopted under the current information society strategy match the identified targets. The State Secretariat for Telecommunications and the Information Society (SETSI) has adopted a comprehensive approach to extending broadband availability despite Spain's relatively low population density and some difficult terrain. PEBA, the national programme to extend broadband to remote and rural areas carried out under the Plan can certainly be considered a success. Spain's broadband availability was as low as 80% of the population in 2004. Thanks to PEBA, it leapfrogged many of its European counterparts and now has to comply with universal service obligations (100% availability).

SETSI has thus identified a clear path to achieving those broadband availability targets and co-ordinated its intervention (infrastructure funding, universal service obligation) with the broader policy and regulatory context, and with measures taken by other institutions. In all key areas of the programme it has been careful to make maximum use of market forces and to minimise any market distortion through their intervention. The funding mechanism used for the Plan is closely related to the one which the European Commission advanced in the context of its proposed Connecting Europe Facility (CEF). This programme is still a proposal and should be discussed at the European Council and Parliament, but in its current form it provides a good indication of the preferred instruments that will make funds available for broadband deployment in Europe in the coming years. Spain proposes a country-based allocation of funds for the CEF.

Globally, therefore, both information society strategies can be considered a success from a network development perspective. Certain aspects remain, however, where Spain can still learn from the experience of other OECD countries. For example, one area that will need ongoing monitoring in the future is network evolution and that the network developments identified can continue to be enabled. Moreover, competition dynamics must be further intensified to promote price-based competition, which would reduce broadband prices and increase penetration.

Framework for NGA investment – the need for a holistic approach

One key foundation of any effort to foster broadband networks must be the recognition that they play an important role in economic and social development. Promoting such networks can be viewed as a jigsaw puzzle in which all the various pieces of the puzzle – relevant policies and drivers – must be put in place. Only then may their availability and use benefit social and economic development.

Once networks are available, the structure of the market is of concern to policy makers, given that take-up, use, innovation, and customer benefits rely on competitive and efficient markets. Competition is crucial to the delivery of better consumer outcomes in terms of price and innovation, which in turn drives technology adoption and take-up rates.

Measures that acknowledge the broad nature of the issues that need to be addressed and do so from both demand-side and supply-side perspectives are likely to deliver far greater benefits than those which concentrate on only one side. Although this report is concerned primarily with network deployment, any valid assessment must also consider the general context and framework under which the telecommunications sector operates.

The CMT has, for example, lowered the price of access to copper loops in recent years (though the most recent price change was an increase), prompting a significant rise in the number of unbundled loops. In turn, prices for bitstream access products afford little margin to the market entrants who rely on them, which provides economic incentives to move to LLU.

Local loop unbundling has enabled investment by entrants

Policy settings in Spain favour third-party access seekers replicating the access network of the incumbent where possible. As such, LLU is the preferred access product for densely populated urban areas, where entrants have built their networks up to the incumbent's local exchange. This, again, does not apply to cable operators, which fully own the infrastructure up to customers' premises. As regards the level of optical fibre deployments in Spain, it remains low (Figure 1.2), albeit in line with major OECD countries like Australia, Canada, France, Germany, and the United Kingdom. However, it lags far behind the OECD leaders, Korea and Japan, and some European economies (the Czech Republic, Denmark, Norway, the Slovak Republic, and Sweden).

Figure 1.2. **Percentage of FTTH connections in total broadband in OECD countries, June 2011**

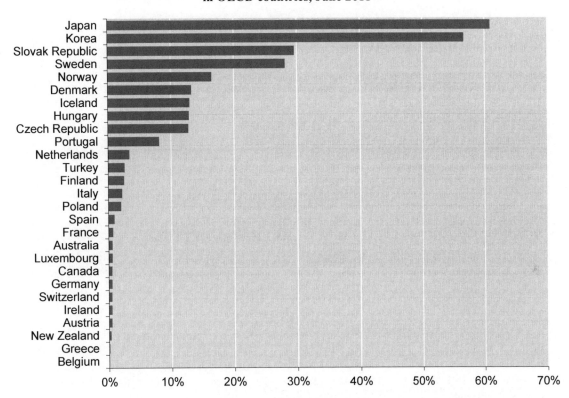

Source: OECD Broadband Portal, www.oecd.org/internet/broadbandandtelecom/oecdbroadbandportal.htm.

Nevertheless, it is clear that alternative operators have reacted positively to the regulatory regime in place and that the migration from bitstream to LLU has gathered pace since 2008, so increasing competition among third-party providers who access Telefónica's telecommunication infrastructure. It should also be noted that Spain is one of the few OECD member countries to have seen the development of a large-scale, end-to-end network provider – ONO. It began operating between 1996 and 1998 and currently serves some 50% of Spanish homes, which makes it Spain's biggest cable provider. The three small regional operators – R, Telecable and Euskaltel – are in a similar situation in their respective regions, although they lack the economies of scale available to ONO, given its larger network (which was recently upgraded to DOCSIS 3.0).

Figure 1.3. **LLU access compared to bitstream and resale, 2005-2010**

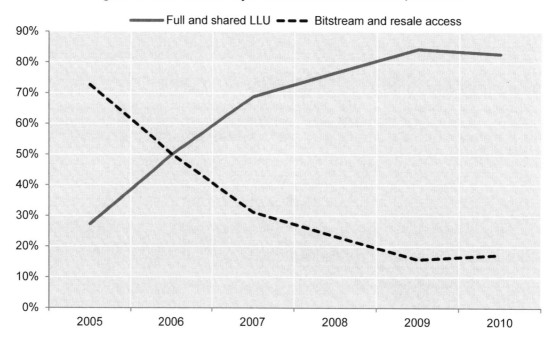

Source: CMT (Comisión del Mercado de las Telecomunicaciones) (2010), *CMT Annual Report 2010*, CMT, Madrid.

Cable operators have, altogether, managed to deploy their own network infrastructure which serves some 60% of the Spanish population. New DSL entrants now (after a few years' time delay) offer a competitive response to Telefónica, due in large part to the increased use of LLU since 2005 (Figure 1.3). They have been steadily taking business from Telefónica (Figure 1.4), whose market share had been relatively stable between 2004 and 2008. The main entrant operators using Telefonica's network are Jazztel, Vodafone and Orange.

Ultimately, investment decisions are made by individuals, institutions, and corporations to serve their own interests, although those interests are influenced by policies which govern the relative costs and benefits of investing in particular areas. In other words, appropriate policies are fundamental in encouraging investment in areas of national interest.

Looking at those countries which have been most and least successful in mobilising investment and increasing their capital formation to ensure the widest available broadband availability suggests that Spain is already in a strong position. A key component of any successful regulatory regime is the creation of adequate incentives and legal certainty for investment in order to deliver NGA networks.

Figure 1.4. **Broadband market shares, 2005-2010**

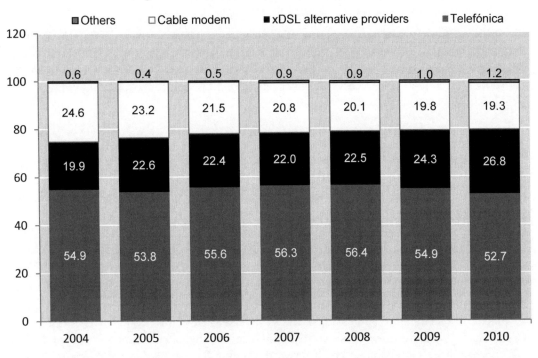

Source: CMT (Comisión del Mercado de las Telecomunicaciones) (2010), *CMT Annual Report 2010*, CMT, Madrid.

Models for the economic development of networks

There are, broadly speaking, three economic policy models for developing networks. In the first model, network development is driven principally by government policy (Cave and Martin, 2010) whose designs supersede investment decisions. A second model may be described as purely market-driven, where public intervention is so limited the authorities may not even regulate market access. A third model lies somewhere between the first two.

The third model fulfils the needs of most consumers and businesses, while public interventions (beyond regulation) are strictly limited to those areas where the private sector would not normally invest due to high risk or perceived low profitability. The Spanish investment framework falls squarely into this third category with competition on and between platforms driving investment in the first instance. However, that investment is influenced by the regulatory framework put in place, which may have contributed to reshaping the market's investment profile over the previous five years.

The broadband market has many demand- and supply-side drivers. On the supply side, the cost of building network infrastructure, for example, may be so high that it undermines the business case. Solutions like sharing passive infrastructure components cut costs and can help to radically lower the cost of deployment. On the other hand, even if the networks are in place, the investment will be realised only when consumers and businesses make use of these networks. That is where demand-side enablers come in. Available on-line services and consumer trust in electronic transactions are as important as successful supply-side measures. There is a clear need to create a virtuous circle between broadband supply and demand, as the OECD (2008a) recognised in its report,

Broadband Growth and Policies in OECD Countries, which assessed the implementation of its recommendations on broadband development (OECD, 2004).

While different bodies – such as SETSI and the CMT – have different, clearly delineated responsibilities when viewed from the supply side, the responsibility for demand-side enabling is spread across virtually all government departments. There is consequently a need for a very high degree of policy co-ordination – including with supply-side measures.

Certainty and incentives for investors

To complement CMT's action to redirect investment, the CMT and SETSI have undertaken a number of facilitative measures which lower deployment costs and, by the same token, investment risks. The most important such measures are spectrum reforms, infrastructure-sharing arrangements, and appropriate financial instruments for deploying broadband networks. From an investment perspective, clearly stated means and ends are deemed to be critical by most observers. In this respect, Spain has performed well, setting clear objectives and policy instruments.

The high level of co-ordination between the different bodies responsible for network regulation also provides legal certainty and assures investors that a coherent framework is in place. A recent study undertaken by Analysis Mason on behalf of the European Commission identifies measures taken within a comprehensive national broadband plan which sets out a clear vision and path forward as being the most effective course of action (Analysys Mason Ltd., 2011).

Providing legal certainty and incentives for investments relies to a great extent on having a stable, predictable regulatory framework that is adequately implemented and enforced. In this respect, the new European e-communications Regulatory Framework was approved in 2009. Member countries were given until May 2011 to undertake the necessary legal changes and transpose the new framework into their domestic legislation. Unfortunately, Spain and 15 other EU member countries have so far failed to do so (European Commission, 2011). Parliament was on the verge of passing the new General Telecommunications Law, but the 2011 general election stopped the process. It should be resumed at the earliest opportunity, so that Spain fully implements the new European regulatory framework on telecommunications.

Regular market reviews

Another area where compliance with current rules should be more firmly enforced is the frequency with which the regulator conducts market analysis and determines where *ex ante* regulation may be warranted. While the new framework states that, with some exceptions, no more than three years should elapse between reviews, the current framework urges national regulatory authorities to carry out market reviews every two years. Spain last carried out most markets reviews between 2008 and 2009. However, the last time it reviewed the market for call origination goes back to 2006, when new conditions for MVNO access were introduced.

Since telecommunication markets change extremely quickly, these compulsory periods should be adhered to. The CMT has already included a new round of market reviews in its programme of work for 2012. A new round must be undertaken at the earliest opportunity and the frequency set out in regulations must be complied with thereafter. This will contribute to increasing legal certainty for all players and help deliver better, more timely regulation.

Universal service reform and the need for a national broadband plan

The European Union's 2009 Telecoms Package has provisions which allow member states to specify not only which services should be "universally available" but which ones should come under the heading of "universal service". The distinction in European Union legislation is important since it leaves member states free to choose which services to deliver. However, the costs of providing services can be borne by industry only if they are deemed to be universal services. Otherwise, the public purse would foot the bill. The EU's 2009 Telecom Package allows Spain the freedom to define the appropriate data rate for network connections delivering "functional Internet access" for the purposes of universal service rules and in the light of national conditions.

Basic broadband connectivity can therefore be part of a universal service at national level in cases where market forces and other policy tools and financing instruments have failed to generate universal broadband coverage. While the European Union considers that, at this stage, it would not be appropriate to mandate broadband at a specific EU-wide data rate, it recognises the right of countries such as Spain to set such rates for themselves.

Broadband affordability and market dynamics

One area of particular concern in the Spanish market is the level of pricing. Figure 1.5 shows the average level of pricing in the OECD for the broadband offer that is most popular in Spain – between 2.5 and 15 Mbps. It indicates that Spain's retail charges are high. While most customers buy bundled offers (not fully reflected in the figure), other studies that take bundles into consideration also support this view (CMT, 2010b).

Some stakeholders have drawn attention to considerable off-price list discounting in the market, which sometimes continues even beyond the original duration of the discount. Such pricing opacity is not, of course, desirable, but attention should be given to take-up rates. They are below the OECD and EU averages. High prices reduce take-up, as the Spanish market demonstrates that the willingness of potential new customers to pay will be determined by the advertised price. Wide discrepancies between advertised and actual prices may have a material impact on take-up rates. Greater price transparency would encourage more users to take up broadband.

While encouraging sufficient competition in the market is certainly CMT's responsibility rather than SETSI's, higher retail pricing is a barrier to take-up and undermines the benefits of the universal coverage work SETSI is co-ordinating. Availability is a necessary but insufficient condition for users to benefit from broadband connections, as promoting take-up and usage are also requirements.

Data indicate that broadband penetration in Spain is slightly less than the OECD average, but also that there is greater scope for improvement (Figure 1.6). Available evidence suggests that access to basic broadband and to more data and information creates opportunities for improving consumer welfare, productivity, and economic growth. Confirmation comes from empirical research across OECD countries which points to broadband uptake having a strong, positive economic impact.

Figure 1.5. **Average monthly subscription (including line charge) for speeds between 2.5 and 15 mbps, September 2010 (USD PPP)**

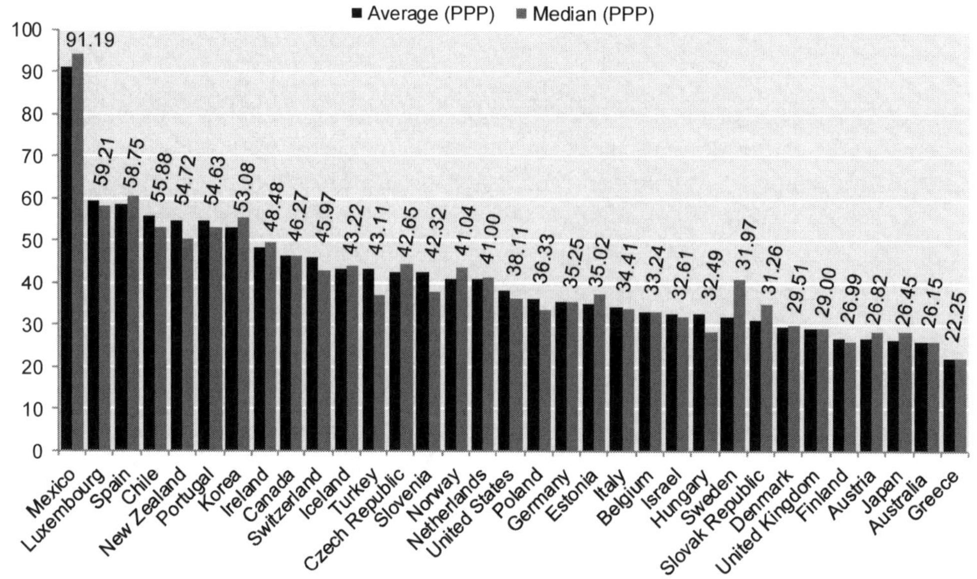

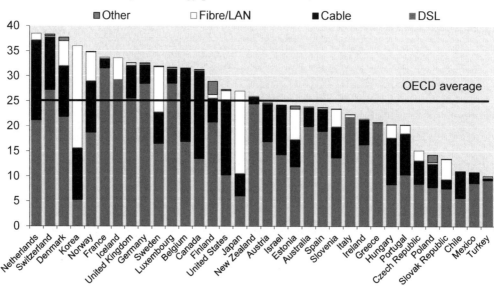

Note: The statistical data for Israel are supplied by and under the responsibility of the relevant Israeli authorities. The use of such data by the OECD is without prejudice to the status of Golan Heights, East Jerusalem and Israeli settlements in the West Bank under the terms of international law.

Source: OECD (2011), *OECD Communications Outlook 2011*, OECD Publishing, Paris, http://dx.doi.org/10.1787/comms_outlook-2011-en; and *OECD Broadband Portal* (2011), www.oecd.org/internet/broadbandandtelecom/oecdbroadbandportal.htm

Figure 1.6. **OECD fixed (wired) broadband subscriptions by technology per 100 inhabitants, June 2011**

Note: The statistical data for Israel are supplied by and under the responsibility of the relevant Israeli authorities. The use of such data by the OECD is without prejudice to the status of Golan Heights, East Jerusalem and Israeli settlements in the West Bank under the terms of international law.

Source: OECD Broadband Portal, www.oecd.org/internet/broadbandandtelecom/oecdbroadbandportal.htm

Broadband and economic growth

Two papers drawing on differences between states in the United States assessed the links between broadband and growth. Although they did not fully resolve the question of causality – does growth drive broadband uptake or does broadband drive growth? – they do find a positive link. Findings from another US-based study did, however, confirm the existence of a causal link between broadband and growth. It also found the impact on growth was stronger in less densely populated areas (Crandall et al., 2007; Gillett et al., 2006).

Another study of 25 OECD countries in the decade to 2007 concludes: "We find that a 10 percentage point increase in broadband penetration raises annual *per capita* growth by 0.9-1.5 percentage points." The study specifically accounts for the separate impact of broadband in addition to those of computers, mobile telephony, and other technologies in use at the time. An EU study further reinforced this point, finding that growth impacts arise from the productivity benefits of broadband use by individual firms.

In the light of evidence that suggests broadband availability and take-up have social and economic benefits, achieving universal availability is desirable. The previous information society strategy greatly increased broadband availability, while the current Plan is now pushing that penetration much further. Figure 1.7 shows the extent of the penetration achieved. Despite the rough terrain in certain rural parts of Spain, broadband now reaches 99% of all households in both urban and rural areas, putting Spain close to the top of the OECD tables (Figure 1.8).

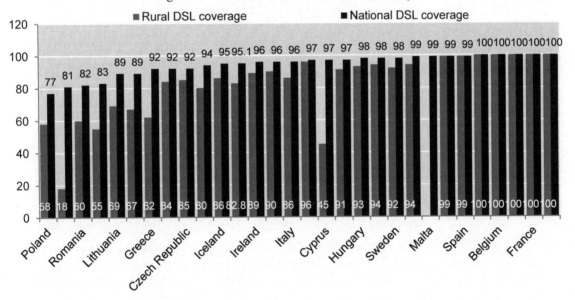

Figure 1.7. **Broadband rural/urban availability**[*]

* Population coverage as percentage of population under a DSL-enabled DSLAM.
Source: IDATE (2010).

However, as noted above, greater emphasis on encouraging retail price competition should be a policy priority in the future in order to ensure that enhanced availability results in increased consumer take-up. This has indeed been the trend in the past two years. If it is to be continued, the Spanish market will certainly benefit from the higher number of consumers able afford broadband services.

Policy sequencing for broadband availability

The Plan's programmes have taken broadband availability to a level where the 100% availability requirement no longer puts an undue burden on industry. From a mid-table position in the OECD broadband coverage ranking in 2005, Spain is in top spot today.

Figure 1.8. **DSL coverage and population density, 2009**[*]

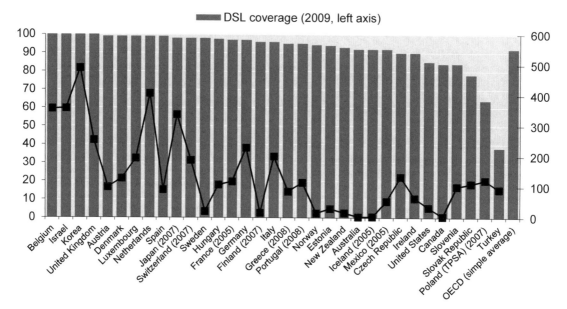

Notes: The statistical data for Israel are supplied by and under the responsibility of the relevant Israeli authorities. The use of such data by the OECD is without prejudice to the status of Golan Heights, East Jerusalem and Israeli settlements in the West Bank under the terms of international law.

* DSL coverage is measured in various ways across the OECD. The percentages given in this figure may represent the number of lines that have been upgraded, the population covered, or the households able to subscribe. Data for the United States is an average for the providers Verizon, SBC, Bell South, Qwest, Sprint, Alltel, Cincinnati Bell, and CenturyTel.

Source: OECD Broadband Portal (2011),*www.oecd.org/internet/broadbandandtelecom/oecdbroadbandportal.htm.*

Furthermore, the Plan Avanza programmes have achieved this rapid improvement in absolute and relative broadband availability without placing an excessive burden on the industry or the state. Instead, they have made co-ordinated use of public funding and wide-ranging spectrum reforms and allocations. These measures were put in place primarily to minimise the number of areas without broadband. The universal service obligation was then enforced, compelling the designated universal service operator (Telefónica) to provide basic broadband connectivity in the remaining areas where there was none. This approach is a good example of policy sequencing. It reduced the net cost to industry of the universal service obligation, as a ministry press release reporting the recent allocation of spectrum noted:

The tender of all the available spectrum, a total of 310 MHz, will allow the provision of mobile broadband in rural areas and the future introduction of the 4th generation mobile telephony (LTE) with the obligation to bring ultrafast mobile broadband coverage to 98% of the population … In this way the spectrum awards will facilitate the fulfilment of the Digital Agenda for Europe's objectives of making basic broadband universally available by 2013 and broadband at ultrafast speeds of 30 Mbps by 2020, while making a powerful contribution to narrowing the digital divide. (Spanish Ministry of Industry, Tourism and Trade, 2011)

There is no one-size-fits-all model for network building. The authorities must consider the pros and cons of each model and how it might fit a particular situation. Because deploying broadband networks requires heavy investment, state support of some kind is often needed in rural areas, while attempts to balance the long-term needs of individuals with the commercial aims of private partners feed into the choice of investment model. Models come in various forms: public investment, private or public-private partnerships like design-build-operate projects, bottom-up projects, joint ventures, and publicly outsourced contracts.

Spain has implemented a very successful strategy in achieving broadband availability throughout its territory. Bearing in mind the country's significant rural population and difficult terrain, the approach has been especially ambitious: broadband has been included in the Universal Service Obligations so that any citizen can request it at the regulated price in line with market rates in urban areas. Spain's Universal Service Obligations (USO) mirrors the objective of the Digital Agenda for Europe:

The overall aim is to deliver sustainable economic and social benefits from a Digital Single Market based on fast and ultra fast Internet and interoperable applications …. to bring basic broadband to all Europeans by 2013 and … ensure that, by 2020: *i)* all Europeans have access to much higher Internet speeds of above 30 Mbps; and *ii)* 50% or more of European households subscribe to Internet connections above 100 Mbps. (European Commission, 2010)

While it is clear that Spain aims to meet the Digital Agenda goals, it has not published any detailed specifications of how it understands those goals or revealed how it intends to meet them. Although it is in the process of meeting the objective of 1 Mbps of universally available broadband, questions remain as to how it plans to meet the 30 Mbps target and ensure that 50% or more households have Internet connections above 100 Mbps by 2020.

Coverage obligations attached to the new spectrum awards will extend LTE coverage to 98% of the population. However, Spain must state its interpretation of the 30 Mbps Digital Agenda target more clearly. There are enormous differences in requirements if this target is interpreted in terms of actual, advertised, or even theoretical speeds (i.e. LTE speeds for customers covered by one base station). Besides, for LTE deployments, even under coverage obligations, there may be a need for additional backhaul infrastructure for serving areas such as small and middle-sized towns. Finally, solutions must be found for the 2% of the population who remain unserved.

Moreover, a 50% subscribership at 100 Mbps calls for a much wider household coverage – 80% or 90%, regardless of competition and affordability concerns. While cable networks now reach 60% of the population, higher coverage would involve extensive fibre deployment by the incumbent (or third-party) operator, significant advances in technology, and greater uptake of wireless technologies.

In short, Spain needs to clarify these issues. One proposal is that it should publish a national broadband plan which sets out targets, draws up a roadmap, and describes existing constraints. Such a plan should also specify what resources are required and whether private investment is sufficient or public funding needed. Any national broadband plan should look carefully at the European context and proposed broadband funding programme like the Connecting Europe Facility.

Finance

The Plan supports network investments by making finance available on favourable terms. The New Avanza Infrastructure Programme, for example, is funded by low-interest loans (Table 1.1). The New Avanza Infrastructure Programme within the Plan should be put into a wider perspective. Spain, a highly decentralised country, provides an important source of broadband network subsidies through programmes sponsored by the regions (*comunidades autónomas*).

Table 1.1. **The New Avanza Infrastructure Programme**
(allocated budget vs. maximum budget in brackets)

Millions EUR

New Avanza infrastructure	Allocated budget 2010 (max)	Allocated budget 2011 (max)	Allocated budget total (max)
Line A (1 Mbps broadband extension)	57.56 (100)	15.31 (50)	72.87 (150)
Line B (ultra-high speed networks > 50 Mbps)	75.36 (100)	16.39 (50)	92.17 (150)
Total	132.92 (200)	31.70 (100)	164.62 (300)

Source: SETSI, (State Secretariat for Telecommunications and the Information Society) (2011), *www.minetur.gob.es/es-ES/GabinetePrensa/NotasPrensa/2011/Paginas/bandaancha090111.aspx* and based on information provided by SETSI.

Loans supported 73 projects in 2010 (around EUR 133 million) and 34 in 2011 (EUR 31 million). The second call for proposals (in 2011) was published on 1 July 2011 and the process finally completed on 18 November 2011. This second call was a continuation of the first one, the main difference being that it was co-funded by the European Regional Development Fund (ERDF). Loan applicants must meet the ERDF's requirements and conditions, with interest rates now being determined on the basis of the so-called "phasing-in/phasing-out regions". The ERDF has, in fact, reduced its budget due to the current economic situation and actually granted only 32% of the available funds in loans in 2011 (Table 1.1). The low application rate is attributable to two sets of factors:

- Interest rates were 4.29% in non-ERDF areas and 1.29% in ERDF areas, up from the 1.428% of 2010. Furthermore, the 2011 call for projects requested some guarantees.

- The overall Spanish economic downturn arguably had the effect of postponing or cancelling investment decisions by operators. Besides, operators had used up a fair amount of their financial resources in spectrum auctions.

Financial instruments and the Connecting Europe Facility

Broadly speaking, the financial options available for broadband network deployment consist of two categories: financial instruments (e.g. risk-sharing initiatives like EU project bonds) and direct grant aid. If broadband projects can be fully funded in the private market, then making that investment cheaper through the use of financial instruments will be attractive to market operators. However, if projects find it difficult to attract private financing in the first place, then the loan terms will be of secondary importance. For such projects, and those that need additional funding, the Connecting Europe Facility (CEF) provides EU backing in the shape of loans, equity, or guarantees (Box 1.1).

If, on the other hand, private-sector loans are not available, then making sufficiently large direct grants available would allow projects to access the balance of capital in the private equity and debt markets. A key issue concerns the extent of any shortfall that public finance would need to make up.

Box 1.1. The Connecting Europe Facility (CEF)

The European Commission has proposed to spend almost EUR 9.2 billion between 2014 and 2020 on pan-European projects to give EU citizens and businesses access to high-speed broadband networks and the services that run on them. The funding is part of the proposed CEF and would complement private investment and public money at local, regional and national level and EU Structural or Cohesion Funds. At least EUR 7 billion would be available for investment in high-speed broadband infrastructure. The Commission considers that this money could leverage a total of between EUR 50-100 billion of public and private investment. In the case of broadband infrastructure, the EU funding from the CEF would leverage other private and public money by giving projects credibility and lowering their risk profiles. The money would be largely in the form of equity, debt or guarantees. This would then attract capital market financing from investors; the Commission and international financial institutions such as the European Investment Bank (EIB) would absorb part of the risk and improve projects' credit rating.

Source: European Union (2011), "Digital Agenda: Commission proposes over €9 billion for broadband investment", press release, Memo 11/709, European Union, 19 October, *http://europa.eu/rapid/pressReleasesAction.do?reference=MEMO/11/709*.

Given the nature of the network projects targeted in information society strategies – providing access or extending backhaul to underserved areas – difficulties in raising finance may be anticipated. In fact, this is often observable in many of the smaller, more localised initiatives which have already received public funds. Direct grant aid can help deliver most of the benefits expected from financial instruments since it supports the business case by lowering the overall level of return on investment required. It also affords private investors peace of mind when choosing whether or not to invest. As a result, the cost of finance and the ease with which funds can be raised is greatly enhanced.

On the other hand, such direct grant aid has a tendency to distort normal market mechanisms and is, in an EU context, subject to state aid rules which ensure that public finance is not used to give competitive advantage to certain operators in EU member countries. Direct grant aid consequently involves longer administrative procedures and a higher degree of uncertainty.

In undertaking any public funding programme for broadband infrastructure, it is equally important to implement an agile decision-making process and a mechanism for the rapid allocation of funds. Allocating funds at the level of the Plan is a serious exercise and calls for an efficiently designed allocation arrangement.

Market failures

Market interventions are normally justified on the grounds that there is some form of market flaw at work. According to Stiglitz (1988), laws fall into the following categories:

- those inherent to the characteristics of the good itself (i.e. public goods, merit goods and externalities);
- those peculiar to the market situation (e.g. flawed competition, which includes information failures and incomplete markets);
- those driven by equity-related factors;
- and finally macroeconomic factors (which affect economic development).

It is clear that in the context of the Plan most, if not all, of these factors are present since there are clear externalities. The markets related to the targeted areas are incomplete, with difficulties in equity of access and digital divide concerns. However, as noted in the section on universal service, basic broadband availability (and indeed NGA) is considered an important enabler of economic development.

An evolving array of relationships between the public and private sectors can be identified and divided into a number of categories (Gomez-Barroso and Feijóo, 2010). Public-private partnerships (PPPs), in the restricted meaning of the term, refer to contractual agreements entered into by a government agency and a private sector entity in which the private sector entity designs, builds, and manages the capital asset and assumes part of the risk (OECD, 2008b). The allocation of funds using the financial instruments in the Plan meet this definition of PPP.

A proposed model (Nucciarelli et al., 2010) for broadband PPPs suggests the need to: *i)* properly identify economic and social targets, *ii)* effectively match the resources and competencies of the different partners; *iii)* design a network in line with the area's geographical constraints (and adopt the most suitable technology solution); and *iv)* define the expected demand and the services required.

Monitoring the information society strategy

It is too early to evaluate the impact of the various initiatives undertaken in the Plan, but any evaluation must consider both the direct and indirect effects. The direct effects are those related to the short-term impact on the penetration of broadband services. These can be measured easily (as long as the metrics are gathered) as the number of connections provided pursuant to a particular PPP project.

The indirect (long-term) effects are much more difficult to foresee and quantify. Some connected people or households might have been able to obtain a broadband connection from another source not provided by public money, although this seems unlikely in view of the targeted areas. A principle difficulty will be the need for more detailed data collection. It will go beyond basic subscription metrics and focus on issues such as the services used and the impact of using them (either in terms of public sector cost savings or benefits to the end user).

While the Plan clearly sets out its intention of putting in place "performance indicators to measure results", available evidence suggests that little data was collected that could be used to judge the impact and performance of investments. And where it was collected, there is no evidence that it was used to assess the impact of the programme. Different metrics could be used for measurement and impact assessment, but the challenge in many instances would be to judge the relative improvement due to the Plan, which is by definition unknown. For example:

- Where networks were deployed using funds, would investment otherwise have been made?
- Where networks were deployed, are they used?
- When used, how were they used?
- What services could users avail themselves of?
- What services did the government deliver differently?
- What was the benefit of such uses?

It is essential to implement effective governance mechanisms to ensure that public money is used appropriately, to check how decisions are made, and to ensure that stakeholders behave in the right way. It is also important to ensure that public money invested in broadband projects is utilised to deliver tangible benefits so that funding continues to be provided only where it is appropriate.

The Plan and, more specifically, the New Avanza Infrastructure Plan, does require operators to file data on the use of the networks that have been deployed using public funds. For example, operators that were awarded funds in 2010 are required to submit a monitoring report every six months starting in July 2012. Some required information is: date of service availability, technical characteristics of the service, the number of new subscribers in those areas where funds were granted, and the number of households connected to NGA networks at speeds higher than 50 Mbps.

While such information filing should continue, there is no evidence that it will be used to monitor the success or failure of the project in terms of economic and social benefits and/or whether any conclusions may be drawn from this information. Nor, to date, has there been any supply- and demand-side cross-checking (e.g. whether an e-government service in a particular area is being used by citizens connected to networks deployed with public funds). There is a clear need for a comprehensive monitoring plan that makes it possible to judge whether the programme has been a success.

Public bodies could usefully undertake periodic monitoring, as they are often the bodies that actually distribute public funds. Local groups and partners could also act as monitors, the advantage being they would be close to operations on the project and thus able to spot any "issues" very quickly. Similarly, if a regional or municipal public organisation were to monitor a programme, it would add financial and political clout to the work.

The disadvantage of public bodies monitoring a project, however, is that they are not directly involved in project implementation. An additional formal process would have to be put in place whereby the operating organisation reports to the regional municipality on a regular basis – as with the Dorsal project in France.[16] That being said, the advantage of a central government body such as SETSI acting as a monitoring body is its awareness of the high-level objectives of any national broadband policy and its close links to the market-specific expertise of the regulator. Public bodies should use milestones and deployment controls to ensure that the roll-out goes according to plan.

To continue with the example of Dorsal, there the network operator specifies to the contracting agency metrics for operational readiness, faults, maintenance schedules, take-up rates and, finally, network performance in terms of speed and quality parameters. These also seem to be important parameters in the context of projects funded under the Plan. Ensuring compliance raises its own concerns and since SETSI does not take a stake in the projects that it finances, other forms of governance that have been effective in other regions/countries could be considered.

In the rural development programme in Sweden (Swedish Ministry of Agriculture, 2008), for instance, most investments are monitored at county level. Investment has to be used as specified in the funding application for a period of five years following the project (monitored by the Department of Agriculture which administers the funds). All projects must meet open-access requirements too, a requirement also made in the previous information society strategy. The Board of Agriculture monitors a 5% sample of projects to ensure that they meet all the relevant conditions. Monitoring requirements are set out in the contract with the network supplier, with payment of funds often conditional on meeting obligations. Such obligations may be launch dates, customers connected, and services provided (normally measured by how many subscribers have signed up with an Internet service provider – ISP).

Finally, governance can be exercised through alternative methods of influence. This approach may be necessary when no formal governance arrangement is possible between the contracting agency and the network operating organisation. However, even in these circumstances, the contracting agency is still able to monitor the project, and refer any undesirable activity to another enforcing body (e.g. the regulator).

In the North Karelia Region in Finland (Regional Council for North Karelia, 2010), the successful bidder for each project has to guarantee 30 years of service provision. One of the North Karelian Regional Council's tasks is to check that the network performance is compliant with the terms of the contract agreement, even though it has no specific tools that it can use to perform this task. As a result, most monitoring is carried out through customer feedback. When problems are spotted, they are signalled to the Finnish regulator (FICORA) to apply penalties where appropriate. With SETSI already responsible for quality-of-service issues and overall consumer complaints on telecommunications, it would make sense that it (rather than the CMT) facilitated customer feedback – and even sanctioned poor performance – in areas where networks have been funded with government support. Such an arrangement would complement the monitoring of other specific measures.

Co-ordination: the "whole-of-government" approach

The previous OECD review of the previous information society strategy identified increased co-ordination as a challenge to which governments must rise before their citizens can reap the dividends of the information society. One proposal would be to ensure closer alignment between supply-side and demand-side measures. For example, e-literacy or e-government programmes could be more tightly integrated with network development measures such as the New Avanza Infrastructure Plan.

There could also be closer co-ordination across regions and within central government itself. Although the institutional framework is already in place, CATSI could be further leveraged to co-ordinate work between regional and local authorities, which could include issues related to policy coherence within central government.

Although most OECD countries need to improve "whole-of-government" co-ordination, there are some successful international examples. Germany hosts an annual ICT summit that brings together federal government, the states (*Länder*) and the private sector. Furthermore, Germany's Federal Network Agency Act has created an advisory council in the Federal Network Agency. It is made up of 16 members of the German *Bundestag* (lower house) and 16 representatives of the German *Bundesrat* (second chamber). The advisory council is consulted regarding spectrum award proceedings, market analysis, universal service obligations, etc., and is entitled to request and obtain information and comments from the Federal Network Agency. The council discusses all ICT-related political issues. Germany also has a states-level ICT working group of the states (*Länderarbeitskreis*), which periodically discusses ICT issues with representatives from the federal government.

In Australia, the Minister for Communications also assists the Prime Minister on digital productivity issues, although this is not seen as a separate function in itself. The title and position are there to assist and emphasise overall co-ordination across the federal government and the economy. They also lend the minister legitimacy and influence in other government portfolios and give him/her a co-ordinating role. The minister is supported by the Digital Productivity Co-ordination Unit run from both his/her own department and the Department of the Prime Minister and Cabinet. The last budget, for example, funded a digital productivity economic package which comprised a range of initiatives across several different portfolios and was co-ordinated by the Digital Productivity Co-ordination Unit.

Spectrum reform

Digital switch-over

The switch-over to digital terrestrial television (DTT) and analogue switch-off was completed in 2010. It was a highly complex process that began with a plan set out in 2005 and a phased switch-off across a total of 90 technical projects. Once DTT coverage and take-up were on a par with analogue television, it was switched off. The switch-off came two years ahead of the required EU 2012 deadline. It was facilitated by SETSI's strong leadership and management throughout the process. Backed by strong national industry support, SETSI helped drive through the extensive legal and regulatory reforms required to enable the switch-over.

Some public funds were used, especially to extend coverage and meet the targets beyond the obligations imposed on broadcasters of 96% private and 98% public television coverage. Public funds were also made available – through tenders conducted by regional governments – to upgrade the extensive regional government networks, which accounted for about one-third of the switch-over's total costs. The process involved issuing invitations to bid for licences. Most were won by terrestrial DTT which had the best cost profile. Significant efforts were also made to ensure that households were also ready for the DTT switch-over.

With the rollout now complete, there are 12 multiplexes serving Spain: 6 private and 2 public multiplexes at national level, and 4 regional and local private and public multiplexes.

Spectrum tenders

In June 2010, SETSI launched a public consultation on several radio spectrum issues in the 800 MHz, 900 MHz, 1 800 MHz and 2.6 GHz bands. It proposed that all spectrum should be assigned in 2011, although 230 MHz would be available only the following year, 60 MHz by 2014, and 20 MHz in February 2015. The assignment of spectrum emerged as a main priority. In 2011, the Spanish authorities assigned 300 MHz through a combination of auction and beauty contest licence award procedures on a technological, service-neutral basis (Table 1.2). The 250 MHz of new spectrum constituted a 70% increase in the spectrum available for mobile services at the time. Spectrum caps had been set at 2x20 MHz below 1 GHz and 115 MHz above 1 GHz.

Table 1.2. **Spectrum tenders – final assignments**

Number of concessions	Band	Block size	Licence award procedure	Winning bidder	Availability
1 national	900 MHz	2 x 5 MHz	Beauty contest	France Telecom	2011-2015
2 national	1 800 MHz	2 x 5 MHz	Beauty contest	Xfera	2011
1 national	1 800 MHz	2 x 4.8 MHz	Beauty contest		
6 national	800 MHz	2 x 5 MHz	Auction	Telefónica, France Telecom, Vodafone	2014
1 national	900 MHz	2 x 5 MHz	Auction	Telefonica	2011-2015
1 national	900 MHz	2 x 4.8 MHz	Auction		2011-2015
4 national	2.6 GHz	2 x 10 MHz	Auction	Telefónica, France Telecom	2011
3 national	2.6 GHz	2 x 5 M	Auction	Vodafone	2011
19 regional	2.6 GHz	2 x 10 MHz	Auction	Ono, Jazztel, R, Telecable, Telecom CLM, Euskaltel	2011 (one vacant block)
19 regional	2.6 GHz	2 x 5 M	Auction	Vodafone	2011
3 national	2.6 GHz	TDD 10 MHz	Auction	France Telecom, Vodafone	2011
19 regional (originally one national)	2.6 GHz	TDD 10 MHz	Auction	Telecable, R. COTA, Euskaltel	2011 (15 vacant blocks)

Source: SETSI (State Secretariat for Telecommunications and the Information Society) (2011), www.minetur.gob.es/es-ES/GabinetePrensa/NotasPrensa/2011/Paginas/nprealdecretoespectro010411.aspx, www.minetur.gob.es/es-ES/GabinetePrensa/NotasPrensa/2011/Paginas/npfinaizacionsubasta010811.aspx, www.minetur.gob.es/es-ES/GabinetePrensa/NotasPrensa/2011/Paginas/npsubastaespectro.aspx

Spectrum in the 2.6 GHz and 1 800 MHz bands and part of the spectrum in the 900 MHz band were also made available. Not all spectrum was assigned. This was partly due to the caps and, more importantly, to the decision of certain operators not to participate at all.

As a result, a new auction[17] was held to award the vacant 50 MHz. A block of frequencies was granted in the 900 MHz band, together with a regional block in Extremadura and five state blocks, both in the 2.6 GHz band. The cap on frequency availability per operator was increased to 25 MHz in the 800 MHz and 900 MHz bands and to 135 MHz in the 1 800 MHz, 2 100 MHz, and 2.6 GHz bands. In order to facilitate proceedings, the block structure was changed to improve compatibility between frequency division duplexing (FDD) and time division duplexing (TDD), while one national block was split into regional ones.

The second auction went on for eight rounds, resulting in four national and four regional blocks being awarded to seven operators: Telefónica (one national block in the 900 MHz band), France Telecom (one national block in the 2.6 GHz band), Vodafone (two national blocks in the 2.6 GHz band), and Telecable, R, Cota and Euskaltel (one regional block each in the 2.6 GHz band). The total public revenue raised was EUR 185.4 million, predominantly from the national blocks – only EUR 800 000 was raised from the regional blocks. Sixteen additional regional blocks have remained vacant.

The previous spectrum sales, which used beauty contest and auction procedures, raised EUR 1 647 million for 51 frequency blocks – 210 MHz of the 250 up for bidding – plus EUR 733 million in investment commitments. Add to that the EUR 168 million from the beauty contests, and nearly EUR 2 billion were raised and 300 MHz awarded.

As discussed in relation to the USO, the Spanish Parliament chose to include broadband in the USO because it believed spectrum reforms and releases made universal broadband feasible. Coverage obligations have been attached to the spectrum awards, requiring a 30 Mbps coverage of 98% of the population by 2020 so that rural and remote areas are served by a high-speed provision. The technology- and service-neutrality of broadband delivery under USO has been increased for the 900 MHz and 1 800 MHz bands (GSM and UMTS technology) and the 800 MHz and 2.6 GHz bands (any technology), which will facilitate LTE deployments in these bands.

Setting the right spectrum caps

The issue of spectrum caps was a bone of contention. The caps concerned were the one below 1 GHz (2 x 20) and the one above (115 MHz, FDD plus TDD), which limits incumbents to 2 x 20 in bands higher than 1 GHz. While some operators felt that lower caps would have been appropriate, SETSI also had to consider that they could undermine operators' ability to deliver the necessary broadband services. SETSI consulted widely with other agencies (e.g. CMT, CNC) and authorities at national and European levels with respect to auction design and particularly the setting of appropriate spectrum caps. After some spectrum blocks were left unassigned in the first auction, capping could have been delayed until players had a clearer view of the market conditions (as the unassigned spectrum blocks may not be used until 2015). Although the result would have been, at that point, a more competitive market structure, SETSI nonetheless chose to prioritise public revenues and finalise the auction before the end of 2011.

Spectrum refarming resolved

A challenge that SETSI had to address was how to free up frequencies for reallocation to broadband communication services. It resolved the issue – spectrum refarming – by linking reallocation to the allocation of new spectrum. SETSI thus generated more levers for itself and so completed refarming ahead of schedule. It addressed potential spectrum competition problems by using spectrum caps for allocations that encouraged more efficient utilisation and allowed efficient competition.

SETSI chose beauty contests to assign certain allocations whose spectrum caps constrained larger operators and were deemed inexpensive. It felt that the use of beauty contests to assign spectrum in those allocations would encourage investment and foster competition. Spectrum in the 1 800 MHz was awarded to the entrant Yoigo – Xfera Móviles – for EUR 42 million and a EUR 300 million investment pledge. Orange – France Télécom – paid EUR 126 million and pledged EUR 433 million for an assignment in the 900 MHz band. The 800 MHz spectrum award also has coverage requirements (98% coverage at 30 Mbps by 2020).

By the end of 2014, the upper frequency range in the UHF band (790-862 MHz) will be freed up and made available for advanced communication services. The national DTT channels, which are currently using that band, will be helped to migrate with the proceeds of the beauty contest.

The beauty contest licence award procedure was well designed and delivered a successful outcome in terms of public revenue, operators' final spectrum holdings, and market structure. There was an attempt to encourage Xfera Móviles (Yoigo) to adopt a more ambitious approach and bid for more spectrum in the lower bands. Even though the attempt failed, the market structure certainly offers a more level playing field for wireless competition: some regional operators will no longer need to join forces with MVNOs to provide wireless services and compete credibly with the big three operators. And although other aspects, such as quantifying the benefits of refarming the 900 MHz and 1 800 MHz bands, could have been undertaken differently, the sheer difficulty of quantifying such benefits should be acknowledged.

> Box 1.2. **Germany moves early to reap the benefits of the digital dividend**
>
> A spectrum auction of some 360 MHz of radio spectrum closed in May 2010 after raising revenue of EUR 4.385 billion. (The 360 MHz were divided into 41 frequency blocks across four frequency bands: 800 MHz, 1.8 GHz, 2 GHz, and 2.6 GHz.) The six applicants and four bidders competed over six weeks and 224 bidding rounds. By far, the most revenue was raised for the upper part of the digital dividend spectrum, the 800 MHz band, because it provides greater coverage, especially in rural areas. Germany was the first country in the European Union to make the 800 MHz band available for mobile broadband.
>
> Spectrum in the 1 800 MHz, 2.0 GHz and 2.6 GHz bands was also auctioned, with Germany's four main MNOs bidding to win more high radio capacity for their mobile broadband services in more densely populated areas. Three of the MNOs obtained spectrum in the 800 MHz band which was capped during the auction. All spectrum made available in the auction will be assigned until 2025.
>
> It should be noted that Germany had a significantly less complex spectrum allocation system ahead of the auction, as its television broadcasting services are less reliant on terrestrial technology. This clearly facilitated the German authorities' early move to auction off spectrum. Spectrum allocation, in countries like Spain and Italy, made the switch to DTT more complex and the release of digital dividend spectrum more difficult.
>
> *Source*: OECD elaboration based on BMWi press release.

As mentioned above, Spain's spectrum allocation to broadcasting and its strong reliance on analogue terrestrial television made its switch to DTT and release of digital dividend a particularly challenging undertaking (Box 1.2). However, SETSI managed to support its wider broadband objectives by using spectrum caps to ensure competition for allocations in high-capacity parts of the spectrum. And although it initially opposed the release of the digital dividend, Spain then acted swiftly and implemented a comprehensive, ambitious set of measures to release digital dividend in the 800 MHz band. Indeed, the Council of Ministers recently approved further new measures to implement the digital dividend.[18] Challenges could arise if additional bands (e.g. 700 MHz) have to be attributed to wireless data services, as current allocations in these bands are equally complex. This is, however, a subject for discussion in the medium term.

One area which will warrant further examination is the extent to which fibre networks will need to be deployed in more remote areas in order to support backhaul. While much work has been done under the Plan, it is likely that further backhaul deployments, deeper in the mobile networks, will be required. While it is up to operators to invest in backhaul deployment, the Spanish authorities should still ensure that there is sufficient backhaul in the wireless networks for public policy objectives to be met.

In conclusion, the spectrum reforms which Spain carried out in 2011 can be considered best practice. They show how political leadership, wide consultation processes, and clear roadmaps can be used to implement change in the wireless industry's most important competitive asset.

Sharing infrastructure and reducing deployment costs

A key source of cost savings in network deployment is the reuse of existing infrastructure. The reuse of existing ducts, for example, avoids the hugely expensive business of digging new trenches to install fibre infrastructure. The EU's Broadband State Aid Guidelines[19] recommend that if a public authority decides to undertake civil engineering work, such as digging new ducts, to enable and accelerate operators' deployment of infrastructure, such public works should not be considered "industry- or sector-specific". In principle, public works should be undertaken for the benefit of potential users (electricity, gas and water utilities, etc.) and not just electronic communications operators.

Other desirable measures include the co-ordination of civil works, mapping, and efficient communication. Streamlined laws and regulations relating to public works, urban planning, environment, and public health should also be in place to enable the smooth acquisition of rights of way and the removal of barriers to the deployment of wireless infrastructure – e.g. difficulties in obtaining permission for new base stations or in renewing contracts for existing ones.[20]

As discussed above (in the section "Rights of way and facility sharing") Spain took two important measures to ensure fair facility sharing in 2008. One compels the incumbent to publish its infrastructure access reference offer, which should be cost-oriented, non-discriminatory and transparent. The other measure, which applies to all providers regardless of their market power, obliges the first operator to reach a building with its fibre network to meet access requests from third parties at reasonable prices and under transparent conditions. SETSI, advised by CADIAU, has also drafted legislation to facilitate network deployment inside buildings and to public works on roads and railways.

Mobile operators have also started to engage in important infrastructure-sharing agreements. The Radio Spectrum Policy Group (RSPG) noted that there are sharing agreements in all EU countries. Some are commercially driven, like the one Orange and Vodafone signed in 2006 for sharing in Spain and the United Kingdom and the 2009 deal between Telefónica and Vodafone governing passive infrastructure in Germany, Ireland, Spain and the United Kingdom.

Sharing agreements may also be encouraged by the competent authorities or required by regulators with particular conditions attached. In Spain, the authorities may mandate the shared use of public and private property when environmental, health and safety, security, and urban planning factors so warrant. In that case, the terms of sharing agreements are left up to the parties, but any dispute that arises can be submitted to the CMT which will set the conditions.

The RSPG's (2011) *Infrastructure Sharing Report* discussed member countries' experience with regulatory and/or government policies as they relate to improving broadband coverage. The indication gleaned from respondents was that, thanks to the ease of establishing new sites, sharing agreements were cutting costs and improving network coverage. The report highlighted as best practice the agreement in Spain between Orange and Vodafone whereby random access network sharing was facilitating the provision of broadband services to small towns with populations of fewer than 25 000.

With respect to in-building wiring, Spanish regulations have undergone a number of makeovers over the decades, each one refining the one before it. Historically, it first passed legislation in the 1960s to allow the installation of communal aerials.[21] More recently, the in-building wiring framework has been fundamentally based on the requirements laid down by a series of royal decrees issued between 1999 and 2006.[22] Since 1999, property developers have been obliged to incorporate these communication infrastructure requirements in new housing developments. Today, roughly 20% of buildings in Spain comply with the requirements laid down in the 1999 royal decree and in further amendments.

The latest change to the in-building wiring framework came in 2011, when the government approved Royal Decree 346/2011.[23] It replaced all previous regulations and updated the framework to accommodate digital terrestrial television, home-networking facilities, and new requirements for coaxial and fibre in-building wiring. The decree also updated the previous regulation to bring it into line with the new digital radio and television standards.

As an option, the 2011 reform contains the elements necessary for deploying home-networking facilities, including a certification system that will allow developers to label new buildings according to their home-networking capabilities. The labelling system has proved to be extremely successful in Korea, where property developers adopted the Building Certification Programme (BCP) to rate apartment buildings according to their capacity to accommodate fibre-based connectivity (Box 1.4).

The main changes Spain adopted in 2011 have brought it closer to the policies advocated in the Digital Agenda for Europe and removed one of the main barriers to infrastructure deployment. The key metrics in the Spanish framework include a trigger for ensuring that basic duct infrastructure is in place, monitoring how investment is financed, and verifying that infrastructure's technical parameters (size, access, and specifications) conform to standards.

While the in-building wiring regime for new buildings is considered excellent, specific concerns regarding the applicability or impact of measures were expressed as the building trade business cycle went off the boil. Some operators consider that only 40% of existing buildings have proper in-building wiring. Compliance with the 1999 framework is estimated to be around 20%, following a decade-long construction boom and three to four years of virtual freeze. Korea's certification system shows that, even with sustained activity in the construction sector, the system takes many years to translate observably into large numbers of properly wired buildings. Nonetheless, it should be acknowledged that Spain has laid the foundations for removing barriers to the deployment of in-building wiring, even though the results may not yet be visible.

Box 1.4. **Building Certification Programme in Korea**

Korea is one of the leading exponents of fibre deployment in the OECD area. The percentage of fibre connections in total broadband subscriptions reached 57% in June 2011, second only to Japan with 61% (OECD Broadband Portal, 2011). When Akamai, a global content distribution network, measured Korea's average Internet connection speed in 2011, it was the fastest it had recorded since it began measuring in 2007 (Akamai, 2011). This successful penetration of fibre networks is, however, underpinned by a decade-long effort to enhance the in-building wiring framework for multi-dwelling buildings in Korea.

The Building Certification Programme (BCP) certifies that an apartment building complex is equipped with suitable communication infrastructure for fibre-based broadband services. For instance, when an apartment building is connected to at least four optical cables (i.e. in-building facilities such as main telecommunication rooms, ducts, and wiring for FTTH services for residents), it qualifies for BCP's supreme grade. BCP has two other grades, first and second, both of which should ensure FTTB connectivity. Unshielded twisted pair (UTP) cables are used for in-building wiring for first- and second-grade buildings and apartments. The programme can be applied to most of the apartments and major buildings in the country. If a building is compliant, the applicant is awarded a certificate from the Korean Communications Commission (see figure below) and given permission to publish and advertise the award.

Korea's Building Certification Programme's certificates

Building Certification Programme Supreme grade Building Certification Programme First-grade
Home network system

Introduced in July 1999, BCP is now the *de facto* standard for in-building wiring, especially in multi-dwelling residential units. The figure below shows the number of apartment units that received a certificate (it almost doubled from 2000 to 2010) and the ratio of certified buildings to apartment blocks built each year. A large number of old apartments also had new in-building wiring facilities fitted to comply with BCP. As of October 2011, 28.8% of Korean households complied with BCP (13.2% with supreme grade, 60.9% with first grade, and 25.9% with the second grade).

> **Box 1.4. Building Certification Programme in Korea** *(cont'd)*
>
> **Apartment buildings in Korea increasingly awarded the BCP certificate, 1999-2010**
>
> *Source*: Korean government (2011)
>
> BCP's early success saw it extended to other communication fields, such as home-networking (356 783 certified units as of October 2011). In September 2011, a DTT reception requirement was added.
>
> The most striking feature of BCP is that it is not based on the regulation of in-house wiring but on competition in the housing market. Current regulations for in-building wiring in Korea do not require facilities for FTTH nor FTTB. Nevertheless, the BCP Programme benefitted from a housing market that expanded at a rate of around 200 000 households per annum, prompting the government to devise an instrument for differentiating housing developments' demand for Internet services (broadband penetration was Korea was already 20% in 2002). The Korean government succeeded in upgrading in-building wiring in most new apartments to the superior fibre-based system. Bearing in mind that 58.6% of Koreans live in apartment buildings, whose infrastructure is costly to upgrade, it may legitimately be argued that Korea has overcome one of the greatest barriers to the deployment of NGA networks.
>
> *Source*: OECD elaboration based on information provided by the Korean government.

Spain will, of course, have to constantly continue updating the framework, just as it has done so far. One specific suggestion, however, is that the general building review – which takes place every 15 years as part of a building's architectural review – could incorporate a requirement to meet the currently specified greenfield standards. In addition, any tax relief for such measures could be highlighted or even extended to making buildings "fibre ready" in the current context of Spain's fiscal consolidation and overall tax policy.

Key assessments and proposals for action

This chapter has dealt chiefly with the supply-side policies that Spain is implementing to boost its network connectivity, especially its access to fixed and mobile broadband services. However, it also considers some desirable elements of alignment with demand-side policies.

Spain's market context and regulatory framework, both heavily influenced by European-wide policies, play a key role in shaping the telecommunications market. They are also crucial in determining the instruments available to the Spanish authorities for taking action on specific issues. The European regulatory framework clearly constrains national policies and practices in areas like the State Aid Framework, procedures for defining and analysing communication markets and, more specifically, remedies for significant market power. Nonetheless, the Spanish authorities – the regulator in the present instance – do enjoy a significant degree of autonomy and play a critical role as well.

In addition to the overall European context, poor economic conditions and ongoing fiscal consolidation currently curb the Spanish authorities' ability to implement more far-reaching public programmes designed to increase network availability and use. The previous information society strategy was among the most ambitious information society strategies undertaken in OECD countries between 2005 and 2009 and dwarfed the resources allocated to the current Plan. As the OECD acknowledged in its first review (OECD, 2010c), the Plan put down very important milestones – increased broadband coverage in rural and remote areas, a successful switch-over to digital terrestrial television (DTT), and significant progress in e-literacy and the availability of online government services.

While not on the same scale as its predecessor, the Plan enjoys a non-negligible budget which, through efficient management and clear policy goals, will certainly continue to improve Spain's performance in the digital economy. Measures taken under the Plan have been successful in further advancing the information society in Spain. This report highlights its achievements and supplies an overview of remaining challenges and action areas that the Spanish authorities should consider for the implementation of measures both now and in the next stages of Spain's information society strategy.

Policy coherence of the Plan

Assessment

The Plan sets infrastructure deployment goals which translate into a series of policy instruments. The outcome is a co-ordinated approach across a number of areas. For example, the objective of universal basic broadband connectivity at 1 Mbps has been met using financial instruments which not only target access networks but aim to put in place the necessary backhaul connectivity. Spectrum policy measures are in line with these connectivity targets, as is the universal service obligation that has been in force since 1 January 2012 and now incorporates the 1 Mbps universal basic broadband connectivity.

The plan is internally coherent and is linked to broader regulatory and policy frameworks, both domestic and European. Measures to remove barriers to the deployment of communication infrastructure like in-building wiring are well designed and contribute to building a comprehensive policy approach towards network deployment in Spain.

> **Key finding**
>
> An important aspect of the Plan is its own policy coherence and its clear connections with policy frameworks, regulatory regimes, and institutional actors.

Network development

Assessment

Spain's approach has been ambitious. **It is one of the few countries to enshrine universal basic broadband availability** in legislation by making 1 Mbps broadband connectivity a universal service obligation. Since 1 January 2012, the designated operator for the provision of universal service, Telefónica, has to meet the obligation of providing functional access to the Internet at 1 Mbps of download speed. The connection may be provided through wired or wireless technologies with a data allowance of 5 GB and a stipulated price.

The inclusion of basic broadband connectivity in USO is the final step in **sound policy sequencing**. Public funding was made available to operators so that they could extend coverage prior to enforcing the broadband USO which, together with spectrum management measures, has substantially reduced the cost of providing universal service from 2012 on.

> **Key finding**
>
> Spain's guarantee of universal access to basic broadband is amongst the most ambitious in the OECD. It met the Digital Agenda for Europe's 2013 target of basic broadband for all by January 2012. Furthermore, the sequencing of its policy to incorporate a broadband service in its USO was well designed.

Assessment

The **other two Digital Agenda goals** are more ambitious and have a much longer time horizon during which technology may evolve and changes in market structures may even occur, making it easier to meet the goals. The Digital Agenda gives neither a detailed specification of the goals nor of how they are to be met.

As regards **the 30 Mbps target**, Spain's spectrum policy measures do appear well co-ordinated. Their aim is to ensure the widest possible coverage through 3G and LTE technologies, which would position it closer to the Digital Agenda targets. However, whether LTE is able to ensure the 30 Mbps target in all scenarios depends on the speed used as a reference (actual, symmetrical, advertised, or theoretical). Moreover, significant backhaul deployments may be needed regardless of the reference speed.

The **100 Mbps 50% household penetration target** is even more ambitious, since it is not likely to be met through wireless technologies alone and will need extensive fibre deployments. Furthermore, "household" penetration requires much higher availability – in the range of 80%-90%.

At the moment, the **incumbent provider Telefónica has deployed fibre only to a limited extent** even though it may have the capacity to deploy extensively in a very short period of time – at least in densely populated areas. For the time being, only cable networks may support speeds of around 100 Mbps and such networks may currently ensure a population/household coverage of only around 60%. This is far lower than would be needed to achieve a 50% household penetration target in 2020: 60% population coverage represents significantly less than 50% household penetration as not all households adopt broadband. And, as cable operators have announced that they are not interested in extending their networks further than the current 60% population/household coverage, the potential for meeting the target of 50% at 100 Mbps hinges on the incentives for the incumbent – or, more unlikely, other DSL entrants – to extend its fibre network to cover 80%-90% of the population.

Wireless technologies should not be ruled out of the running for meeting the 100 Mbps target. In rural and remote areas, at least, evolved LTE and related technologies, which will be deployed up to 2020, may well meet needs.

Key findings

Spain has not yet specified in detail how it intends to meet the broadband connectivity targets of the European Agenda for Europe. The ability of wireless solutions to meet the criteria will depend on how the criteria are defined. But under any definition, other enabling measures – particularly with respect to backhaul infrastructure – are likely to be needed.

It is not clear how Spain will meet the second and third Digital Agenda targets. Whether it does so or not depends on: *i)* the incentives for the incumbent to undertake large-scale deployment of fibre networks; *ii)* the extension of cable's footprint to far wider coverage than the current 60% of the population; *iii)* the evolution of wireless technologies; and *iv)* the feasibility of large-scale public funding of broadband networks in the event that market forces fail to meet the targets.

Recommendation

Spain needs to formulate a National Broadband Plan which specifies how, when, and on what assumptions it will meet the Digital Agenda and/or national connectivity targets. Such a plan should state, for example: *i)* the specific quality of service requirements for the targets; *ii)* the technologies able to meet the requirements, which includes connectivity needs at the access, backhaul and backbone levels; *iii)* whether private operators are expected to rely on themselves to achieve the goals or whether public funding will be forthcoming.

Ensuring legal certainty and investment incentives for NGA networks

Assessment

As noted earlier, guaranteeing a **stable framework for investment and legal certainty** is like putting together a jigsaw puzzle. The different pieces are drivers that will eventually form adequate incentives for players to invest in NGA networks. Among the many drivers and factors that should interact to deliver a successful outcome, the following may be distinguished: demand-side measures like widely available e-government services; enablers that ensure a sense of security and trust in networks (e.g. the electronic identity card initiative); and other instruments that lower deployment barriers such as infrastructure sharing and access to ducts and conduits.

Finally, **pro-competition regulation should always be in place**, as the benefits delivered by competition are crucial to ensure long-term market efficiency. Furthermore, most supply-side policies related to NGA networks are determined to some extent by European policy, either through legislation or related provisions such as the NGA recommendation.

Within the European e-communications regulatory framework, Spain has managed to forge an **institutional arrangement** that affords players in the communications sector sufficient legal certainty. Co-ordination between the different public bodies (SETSI, CMT, CNC) has been satisfactory and the current framework provides adequate degree of regulatory certainty. SETSI formulates policy, CMT applies ex-ante regulation and CNC promotes competition and undertakes antitrust and merger review, together with the European Commission.

> **Key finding**
>
> From the point of view of supply-side measures, the current framework provides a sufficient degree of co-ordination and legal certainty.

Assessment

There has been a **significant delay in amending the current regulatory framework**. Spain should have transposed the new electronic communications directives into national law by May 2011 and the passage of the General Telecommunications Bill through Parliament was halted by a general election in November 2011.

Similarly, **the regulator has not yet undertaken the obligatory biennial market review** (triennial under the new regulatory framework). Some markets, like former Market 15 (call origination in public mobile telephone networks) have not been reviewed since 2006. As the CMT conducted its most recent market reviews in 2008 and 2009, the two-year period has already elapsed.

> **Recommendation**
>
> Spain should direct its efforts towards transposing the new European E-communications Regulatory Framework as rapidly as possible. The regulator should, moreover, undertake a new round of market reviews in order to: *i)* comply with the European regulatory framework; *ii)* take due account of the latest developments in e-communications markets; *iii)* adopt the measures necessary to promote competition; and *iv)* increase legal certainty for industry players and other stakeholders.

Assessment

Despite the falling availability of funds and the limited use of available loans under the Plan, financial instruments in the form of **soft loans** are perceived as being an effective means on the supply side (Parcu et al., 2011). Although discussion is still ongoing, significant European funds are likely to be made available in the coming years under the Connecting Europe Facility (CEF) and associated financial instruments. Spain should benefit from these programmes in the near future in order to meet the Digital Agenda targets.

> **Key finding**
>
> The use of financial instruments under the Plan for broadband infrastructure funding is in line with the European Commission's funding proposal. These instruments are also seen as minimising market distortions caused by public intervention and maximise resource mobilisation.
>
> **Recommendation**
>
> Spain should maximise the benefits derived from using European Funds (under the future CEF) to leverage private investment in broadband infrastructure networks, as the ambitious goals set by the Digital Agenda can benefit from additional public funding.

Co-ordinating supply- and demand-side measures

Assessment

While inter-institutional co-ordination works well for supply-side measures only (between SETSI the CMT and the CNC), the present report has identified **gaps in the interaction between demand-side and supply-side policies**. This perception echoes the OECD's in its first review of the previous information society Strategy (OECD, 2010c), where it pinpointed improved horizontal and vertical co-ordination as one of the remaining challenges. By way of example, there seems to be no match between geographical areas to which funds for broadband infrastructure deployment have been allocated and geographical allocations for demand-side measures like e-government and e-literacy.

> **Recommendation**
>
> Efforts should be made to co-ordinate supply- and demand-side measures more closely. One possible area for improvement is the geographical alignment of broadband infrastructure funding and such demand-side measures as e-literacy or e-government programmes.

Assessment

Increased co-ordination may happen in a number of ways. Some were already highlighted in the first OECD review (OECD, 2010a). **Multi-level governance could be strengthened by highlighting the role of CATSI**, Spain's Telecommunications and Information Society Advisory Board. Regional and local governments are represented on this advisory body but their role could be enhanced. CATSI could co-ordinate the alignment between infrastructure deployment measures and demand-side policies through a more effective co-operation scheme between different levels of government. Such increased co-ordination would be required not only across the different levels of government, but also within central government itself in areas like e-literacy or e-government services.

> **Recommendation**
>
> Increased co-ordination should be conducted across local and regional government and within the central government.

Monitoring the benefits of broadband funding programmes

Assessment

Any broadband funding programme should incorporate a mechanism for conducting **sound cost-benefit analysis**. While the costs of the New Avanza Infrastructure Plan may be easily quantified, it is certainly harder to account for the social and economic benefits associated with network infrastructure deployments that have been funded.

As laid down in the requirements for calls for tender, the plan collects information on the technical characteristics of the networks that have been deployed using public funds and subscribership rates once the networks are in place. This data is already being collected and processed internally. However, searching for evidence on the actual use of the networks, alignment with demand-side policies, and the economic and social benefits that these networks enable is far more difficult. For example, one area that could be looked at is the type of applications or services that consumers and business use – e.g. e-government services, teleworking, e-health, e-learning, information purposes only. Even though it is difficult to quantify some in economic or social terms, it is crucial that SETSI obtain a clear picture of the actual use of subsidised networks.

> **Recommendation**
>
> Spain should undertake a thorough assessment of the economic and social benefits brought about by publicly funded broadband networks – e.g. new businesses created, new business models, new learning possibilities, e-health services. It should also perform a sound cost-benefit analysis.

Affordable broadband prices and penetration rates

Assessment

Having a competitive broadband market with affordable prices for consumers and businesses does not fall directly within the scope of the previous information society strategy or SETSI's responsibilities. Nevertheless, such goals are as crucial as other measures in ensuring that citizens and business reap the benefits of the information society. In this respect, **broadband penetration in Spain lags behind EU and OECD averages** and its price levels are, according to available evidence, relatively high.

The incumbent's broadband market share, which remained stable for many years, has been falling since 2009. This trend may be due to more effective enforcement of LLU remedies, the stronger competitive position of DSL entrants, and/or increased consumer price sensitivity in the current difficult economic conditions. Price opacity in the market remains nonetheless of concern. It is reported that considerable off-price-list discounting takes place. It targets specific users – especially those who threaten to terminate their contracts – and discounts often remain in place even beyond their original duration.

While some may argue that these are legitimate commercial practices, they do make it more difficult for consumers to compare prices, while the perception of high prices, whether true or not, discourages potential new subscribers.

> **Key finding**
>
> A competitive broadband market and affordable prices are crucial to ensure take-up. Spain lags behind its OECD and European counterparts in both areas, even though the situation is improving.
>
> **Recommendation**
>
> The Spanish regulator should continue to encourage price-based competition in the broadband market. It should also make competition dynamics more transparent for consumers and address off-price-list discounting.

Spectrum policy

Assessment

Spain started the digital switch from a particularly difficult position due to its historically intensive use of terrestrial broadcasting technology.

New **spectrum holdings were assigned using a well-designed and executed licence** award procedure which set appropriate spectrum caps. Capping was designed to facilitate the entry of new spectrum holders (like Jazztel or the cable operators) and to encourage Yoigo, the fourth mobile network operator to increase its spectrum holdings. Historical spectrum allocations may, in light of technological and commercial developments, need to be recalibrated to facilitate service development and competition both now and in the medium term.

The **use of caps and, where appropriate, voluntary spectrum refarming, appears to have found the right balance in Spain**. Overall, the spectrum allocations comply with service- and technology-neutral requirement and allow LTE deployment in the 800 MHz and 2.6 GHz bands. Coverage obligations have been also introduced to provide 98% of the population with 30 Mbps connectivity by 2020 (by whatever technology is the most efficient at the time).

Some auction outcomes were not anticipated: **some spectrum was not assigned in the first round**, for example, after Yoigo chose not to take part. The second auction could have also been delayed until the perception of market conditions or the economic situation improved. However, the final outcome did address all knowable factors, met the need for additional public revenues in a parlous economic situation, and gave momentum to spectrum allocation reform. And, in a short period of time, it established in a stable, spectrum policy landscape that will be in place until 2030.

> **Key finding**
>
> Spain has achieved a balanced outcome in terms of spectrum policy, final spectrum holdings of all operators, public revenue and market structure. Further involvement of smaller players would have been desirable and was facilitated even if ultimately it did not happen. Coverage obligations have also been introduced, which will assist in providing connectivity to rural areas.

Assessment

The **transition to DTT has been successful** so far. Nevertheless, Spain started to release digital dividend only in 2012, with SETSI having prepared an ambitious set of measures aimed at the 2015 target. Challenges remain: *i)* there needs to be a second migration to free up the 800 MHz band currently occupied by some DTT channels; *ii)* intensive use of the neighbouring bands for broadcasting purposes makes the risk of harmful interference higher than in other countries.

Another possible long-term concern is the possibility of an international agreement that would allocate other bands (e.g. 700 MHz) to wireless communications. Again, the reason for this would be the heavy use of this band for broadcasting services.

> **Key finding**
>
> Spain has so far completed the transition towards DTT successfully. Its last spectrum auction enabled it to lay the foundations for digital dividend release. Nevertheless, digital dividend release ahead of 2015 still poses significant challenges.
>
> **Recommendation**
>
> Spain should move forward and ensure the implementation of the Digital Dividend.

In-building wiring and infrastructure sharing arrangements

Assessment

Spain has made a **sustained, continuous effort to facilitate the deployment of communication infrastructure in multi-dwelling buildings** through measures for in-building wiring. Given the large proportion of residential multi-dwelling units in Spain, any measure that paves the way for the deployment of communication infrastructure in such buildings will certainly remove one of the main barriers to ICT uptake.

These measures have been successful and now some **20% of Spanish buildings comply with the requirements of the 1999 CTI regulation**. The adoption rate is nonetheless slow and heavily reliant on activity in the construction sector. In this respect, the adoption of the new 2011 regulation is likely to be slow, given current levels of activity in the construction sector.

Key finding

The new in-building wiring regime for new buildings is considered excellent and will certainly help to remove barriers to infrastructure deployment in apartment buildings. Given the currently low levels of activity in the construction sector, the effects of the new regulation are likely to be felt only in the very long term.

Spain has put in place a number of measures to facilitate infrastructure sharing, such as the symmetrical obligation on all operators to ensure other operators access to in-building fibre infrastructure at reasonable prices and under transparent conditions. Another example is the proposed obligation laid down by law that requires infrastructure developers to make ducts, conduits, masts and land space available for the deployment of fibre and mobile communication infrastructure in newly built railway and road infrastructure.

Assessment

There are possible ways to **boost the effectiveness of the new CTI** (Common Telecommunication Infrastructure) **regulation** and which may, to some extent, be applicable to existing buildings (which do not fall within the scope of the regulation). Possible tax relief related to the construction, transfer, or renovation of buildings could be highlighted or extended to fibre-ready buildings. Labelling schemes, similar to the one in place in Korea, could be effective in this respect. Furthermore, Spain should consider the possibility of incorporating compliance with new CTI regulation into the general building review which takes place every 15 years.

Recommendation

To make the new CTI regulation even more effective, Spain should consider measures like extending or highlighting existing tax reliefs to cover fibre-ready buildings and introducing labelling schemes for such buildings. It should also consider incorporating CTI compliance in the general building review which takes place every 15 years.

Notes

1. "Telecommunications", "communications" and "electronic communications" are used in this report with the same meaning. "Electronic communications" is the term used in European Law.

2. The new regulation (Royal Decree 346/2011) was approved by the Council of Ministers in March 2011.

3. Telecommunications bill website:
   www.lamoncloa.gob.es/consejodeministros/referencias/_2011/refc20110513.htm#Telecomunicaciones.

4. Real Decreto 458/2011, de 1 de abril, sobre acutaciones en material de espectro radioeléctrico para el desarrollo de la sociedad digital (Royal Decree 458/2011 of 1 April 2011 relative to action in the field of radioelectric spectrum to further the digital society), *www.boe.es/boe/dias/2011/04/02/pdfs/BOE-A-2011-5936.pdf.*

5. Law 2/2011 of 4 March on the Sustainable Economy,
   www.boe.es/boe/dias/2011/03/05/pdfs/BOE-A-2011-4117.pdf.

6. It should be noted that a far-reaching reform of the CMT, alongside with other regulatory bodies, is envisaged by the current Spanish Government. As announced on 24 February 2012, a draft bill proposes to merge all regulatory bodies, creating a single, multi-sectorial, regulatory authority. The new authority, named the National Markets and Competition Commission (CNMC), will merge the current Spanish Competition Authority (CNC) together with other seven sector-specific regulatory bodies. Among those, one of the most relevant is the Telecommunications Market Commission (CMT), as is the energy regulator (CNE). According to the draft bill, the new authority would be related to the Spanish Ministry of Economy and Competitiveness, whereas the CMT was hitherto linked to the Ministry of Industry, Energy and Tourism. These changes were proposed after finalising this report and are therefore not covered in the text.

7. Prior to December 2011 this ministry was the Ministry of Industry, Tourism and Trade.

8. The European Framework for Electronic Communications ("Telecoms Package") was approved in 2002 then transposed into national law in 2003. The revised framework, passed in 2009, must be incorporated into domestic legislation by May 2011. The Spanish 2003 General Telecommunications Law is in the process of being amended accordingly.

9. Nuevo Servicio Ethernet de Banda Ancha (NEBA) (New Ethernet Broadband Service), *www.cmt.es/detalle-oferta-neba?p_p_id=101_INSTANCE_1Two&p_p_lifecycle=0&p_p_state=normal&p_p_mode=view&p_p_col_id=column-3&p_p_col_count=1&_101_INSTANCE_1Two_struts_action=%2fasset_publisher%*

2fview_content&_101_INSTANCE_1Two_urlTitle=111116_neba&_101_INSTANCE_1Two_type=content&redirect=%2fvigente-neba

10. See Case ES/2008/0820: imposition of symmetrical obligations on electronic communications operators with regard to in-house wiring for the deployment of NGA access, pursuant to Article 12 of Directive 2002/21/EC.

11. Infraestructura Común de Telecomunicaciones (Common Telecommunications Infrastructure [ICT]) approved by Royal Decree 346/2011.

12. The Royal Decree on ICT rules was approved in March 2011 and the work continues regarding the other draft piece of legislation on roads and railways (additional clause nº5, Law 56/2007 on Measures to Promote the Information Society).

13. *http://eur-lex.europa.eu/LexUriServ/LexUriServ.do?uri=CELEX:52010DC0245R(01):EN:NOT.*

14. *http://ec.europa.eu/europe2020/index_en.htm.*

15. *www.minetur.gob.es/es-es/gabineteprensa/notasprensa/documents/declaracióncastellano.pdf*

16. Dorsal is a collective project aimed at improving broadband access in the Limousin region (France). It provides broadband service at affordable prices, under uniform tariffs (same for rural or urban areas), and attempts to foster innovation and ensure the development of the region. See *www.dorsal.fr/v3/index.php.*

17. Orden ITC/1074/2011 de 28 de abril (Order ITC/1074/2011 of 28 April 2011) *www.boe.es/boe/dias/2011/04/29/pdfs/BOE-A-2011-7683.pdf.*

18. *www.mityc.es/es-es/gabineteprensa/notasprensa/documents/npplandividentodigital181111.pdf.*

19. Community Guidelines for the application of state aid rules in relation to the rapid deployment of broadband networks, available at *www.broadband-europe.eu/Pages/DocumentDetail.aspx?ItemID=72.*

20. Poland Act on the Development of Telecommunications Networks and Services.

21. Ley 49/1996, de 23 de julio, sobre antenas colectivas (Law 49/1966 of 23 July 1996 on Communal Aerials).

22. Royal Decree 279/1999 (*www.boe.es/boe/dias/1999/03/09/pdfs/A09207-09242.pdf*) replaced by Royal Decree 401/2003 in 2003, *www.boe.es/boe/dias/2003/05/14/pdfs/A18459-18502.pdf*) and amended in 2006.

23. *www.boe.es/boe/dias/2011/04/01/pdfs/BOE-A-2011-5834.pdf.*

Bibliography

Akamai (2011), "State of the Internet", quarterly report, Akamai Technologies, *www.akamai.com/stateoftheinternet*.

Analysys Mason (2011), "Guide to broadband investment", report for the European Commission, 18800-426a, October, Analysys Mason Ltd., London.

Cave, M. and I. Martin (2010), "Motives and means for public investment in nationwide next generation networks", *Telecommunications Policy*, 34(9), Elesevier.

CMT (Comisión del Mercado de las Telecomunicaciones) (2010a), *CMT Annual Report 2010*, CMT, Madrid, *http://informeanual.cmt.es/docs/INFORME%20ANUAL %20CMT%202010.pdf*.

CMT (2010b), Comparativa internacional de ofertas comerciales de banda ancha en la Unión Europea y España, CMT, Madrid, *www.cmt.es/es/publicaciones/anexos/101229 _InformeBA_Junio2010.pdf*.

CMT (2011), *CMT Monthly Report September 2011*, CMT, Madrid, *www.cmt.es/es/publicaciones/anexos/NOTA_MENSUAL_SEPTIEMBRE_2011.pdf*.

Council of the European Union (2010), "Council conclusions on Digital Agenda for Europe", 3 017th Transport, Telecommunication and Energy Council meeting, Brussels, 31 May 2010.

Crandall, R., W. Lehr and R. Litan (2007), "The effects of broadband deployment on output and employment: a cross-sectional analysis of U.S. data", *Issues in Economic Policy*, 6, The Brookings Institution, Washington, D.C.

European Commission (2010), *A Digital Agenda for Europe*, European Commission, Brussels.

European Commission (2011), "Digital Agenda: Commission presses 16 member states to implement new EU telecoms rules", *Europa Press Releases*, IP/11/1429, European Commission, Brussels.

European Union (2011), "Digital Agenda: Commission proposes over €9 billion for broadband investment", press release, memo 11/709, European Union, 19 October, *http://europa.eu/rapid/pressReleasesAction.do?reference=MEMO/11/709*.

Gillett, S. et al. (2006), "Measuring the economic impact of broadband deployment", Final Report, National Technical Assistance, Training, Research and Valuation Project, #99-07-13829.

Gomez-Barroso, J. and C. Feijóo (2010), "A conceptual framework for public-private interplay in the telecommunications sector", *Telecommunications Policy*, 34(9), Elsevier.

IDATE (2010), "Broadband Coverage in Europe Final Report 2011 Survey Data as of 31 December 2010", IDATE, DG INFSO 80106 C December 2011, http://ec.europa.eu/information_society/digital-agenda/scoreboard/docs/pillar/broadband_coverage_2010.pdf

Nucciarelli A., B. Sadowski and P. Achard (2010), "Emerging models of public-private interplay for European broadband access: evidence from the Netherlands and Italy", *Telecommunications Policy*, 34(9), Elsevier.

OECD (2004), "Recommendation of the OECD Council on Broadband Development", adopted by the Council 12 February, OECD, Paris.

OECD (2008a), *Broadband Growth and Policies in OECD Countries*, OECD Publishing, Paris, *http://dx.doi.org/10.1787/9789264046764-en*.

OECD, (2008b), *Public-private Partnerships: In Pursuit of Risk-sharing and Value for Money*, OECD Publishing, Paris, *http://dx.doi.org/10.1787/9789264046733-en*.

OECD (2010a), *Good Governance for Digital Policies: How to Get the Most Out of ICT: The Case of Spain's Plan Avanza*, OECD Information Society Reviews, OECD Publishing, Paris, *http://dx.doi.org/10.1787/9789264031104-en*.

OECD (2010b), *OECD Information Technology Outlook 2010*, OECD Publishing, Paris, *http://dx.doi.org/10.1787/it_outlook-2010-en*.

OECD (2011), *OECD Communications Outlook 2011*, OECD Publishing, Paris, *http://dx.doi.org/10.1787/comms_outlook-2011-en*.

OECD Broadband Portal (2011), *www.oecd.org/internet/broadbandandtelecom/oecdbroadbandportal.htm*

Parcu, L. et al., (2011), "Broadband diffusion, policy and drivers", Florence School of Regulation, Telecommunications and Media.

Radio Spectrum Policy Group (RSPG) and Body of European Regulators for Electronic Communications (BEREC) (2011), "Report on infrastructure and spectrum sharing in mobile wireless network", June.

Regional Council for North Karelia (2010), "Best practices from North Karelia: towards the Europe 2020 Strategy", Regional Council for North Karelia, Finland, *http://ita-suomi.fi/alueportaali/www/fi/aineistoja/Bestpractices_North_Karelia_EU2020.pdf*.

SETSI (State Secretariat for Telecommunications and the Information Society) (2011), "El Ministerio de Industria concede ayudas de 133 millones para el desarrollo de la Banda Ancha", *www.minetur.gob.es/es-ES/GabinetePrensa/NotasPrensa/2011/Paginas/bandaancha090111.aspx*, 9 January.

SETSI (State Secretariat for Telecommunications and the Information Society) (2011), "El Gobierno aprueba el Real Decreto que abre la licitación por subasta y concurso de 310 MHz de espectro radioeléctrico", press reléase, *www.minetur.gob.es/es-ES/GabinetePrensa/NotasPrensa/2011/Paginas/nprealdecretoespectro010411.aspx.*, 1 April.

SETSI (State Secretariat for Telecommunications and the Information Society) (2011), "La licitación del espectro consigue 1.815 millones para el Tesoro Público", press reléase, *www.minetur.gob.es/es-ES/GabinetePrensa/NotasPrensa/2011/Paginas/npfinaizacionsubasta010811.aspx*, 1 August.

SETSI (State Secretariat for Telecommunications and the Information Society) (2011), "El Gobierno cierra la licitación del espectro radioeléctrico con unos ingresos de 2.000 millones para el Tesoro Público", *www.minetur.gob.es/es-ES/GabinetePrensa/NotasPrensa/2011/Paginas/npsubastaespectro.aspx*, 10 November.

Spanish Ministry of Industry, Tourism and Trade (2011), "The Ministry of Industry begins the bidding process to complete the awarding of the entire radio spectrum", press release on Meeting of the Telecommunications and IS Advisory Council, Ministry of Industry, Tourism and Trade, Madrid, 9 May.

Stiglitz, J.E. (1988), *Economics of the Public Sector*, W.W. Norton & Company, New York.

Swedish Ministry of Agriculture (2008), "Rural Development Programme for Sweden – the period 2007-2013", Swedish Ministry of Agriculture, Stockholm, *www.regeringen.se/content/1/c6/08/27/27/ee703769.pdf*.

Telefónica (2011), "Oferta de Referencia de Accesso a los Conductos y Registros de Telefónica" (Reference offer for access to Telefónica's ducts and poles), Telefónica, November, *ww.cmt.es/cmt_ptl_ext/SelectOption.do?nav=oferta_marco&detalles=09002719800aedbb&pagina=1&categoria=vi*.

Chapter 2

E-government: Reforming through information and communication technologies

This chapter covers the Spanish government's overall efforts to develop a paperless administration as well as the specific initiatives to foster digitisation in the two domains of taxation and justice. The chapter covers first the overall framework of the Plan relevant to the fostering of paperless administration followed by the presentation of the Spanish e-government setup within this context. Accordingly, the e-government agenda and the digitisation of the specific areas of taxation and justice will be reviewed, focusing on strategy and co-ordination; digital service supply and channel strategy; digital service demand and uptake; and, finally, consolidation and realisation of benefits. A benchmarking of the Spanish achievements based on available key performance indicators is included when relevant, as well as a selection of international good practices that could inspire and support the Spanish effort to foster further e-government development and implementation. The chapter concludes by summing up key assessments and proposals for action.

The Spanish government faces a series of serious challenges to the economic development of the country, badly hit by the global financial and economic crisis. Increasing trust in the government administration and its policies, reducing expenditures and ensuring efficiency are the main preoccupations. The government's use of ICTs can be supportive in this regard.

This chapter examines Spain's progress in e-government towards the ultimate goal of paperless administration by 2015 – one of the Plan's prime objectives. At the request of the Spanish government, this chapter does not cover the Plan as a whole, but focuses on the progress it has accomplished towards the ultimate aim of paperless administration. Furthermore, the study does not examine general Spanish e-government ambitions, but specifically examines the digitisation of the justice and tax systems. However, to put its analysis in context, it does consider the overall e-government approach.

E-government – towards a paperless public administration

The Plan was approved by the Spanish government in a context affected by the impact of the economic and financial crisis. The challenges of facilitating economic growth and improving long-term employment and competitiveness contributed to an increased focus on ICTs (OECD, 2010d).

The Plan aims to revitalise the economy by contributing to an economic paradigm shift based on the idea of a virtuous circle of productivity and improved public and the private sector ICT capabilities (OECD, 2010d). In respect of e-government, the hoped-for outcomes are, in particular, an increase in quality and user satisfaction (benefiting both citizens and businesses). Cost reductions through reductions of inefficiencies, better procurement prices, and the elimination of duplicated investments are also targets. Although the digital economy shows some signs of recovery, the ICT share of the value added in the business sector and the contribution of ICT-related employment in the Spanish economy are still below the EU average (OECD, 2010a; OECD, 2011b).

The concrete e-government objectives

The ten objectives set out in the plan are linked to e-government in that they seek to increase citizens' and businesses' use of and trust in ICTs, promote the role of ICTs in greening the economy, and generally develop the ICTs and digital content sector (Chapter 1) (Spanish Ministry of Industry, Tourism and Trade, 2010a).

Only the first three, however, relate to the ambition of achieving paperless administration by 2015 (Box 2.1). Of those, only the first one specifically concerns e-government goals. This review addresses that goal. It does not examine the second and third – healthcare and education and the social services.

The concrete implementation of the objectives could take on different directions and levels of ambition depending on the more specific implementation strategy to be chosen. While the Plan's strategy document sets goals for the digitization of healthcare, education and the social services, it does not do so for the e-administration of justice or taxation. The introduction of ICTs into the judicial system is included in the Plan as part of the effort to develop public services, thereby sustaining the Ministry of Justice's 2009-12 modernisation strategy.

> **Box 2.1. Achieving a paperless administration in 2015**
>
> Promoting ICT innovation processes in public administration bodies to put them at the service of citizens and enterprises. The advantages of e-government is a proven fact. It is therefore necessary to encourage the use of advanced services by citizens and enterprises and make public administration receptive to the needs of SMEs and citizens.
>
> - In 2015, a paperless administration will be achieved through the digitisation of procedures, the incorporation of electronic signatures (electronic ID) and the establishment of electronic records.
>
> - In 2013, savings of 3% in the computer resources of public administration through the sharing of resources will be achieved. These savings will reach 5% in 2015.
>
> - For 2015, 50% of citizens and 90% of enterprises will communicate with the administration via the Internet.
>
> *Source*: Spanish Ministry of Industry, Tourism and Trade (2010), *Estrategia 2011-2015, Plan Avanza 2: Anexos*, Ministry of Industry, Tourism and Trade, Madrid.

The criteria for incorporating some areas of e-government service and not others in the Plan do not seem fully clear as indicated in the examples above. The partial inclusion of policy domains could well reflect the fact that targeted efforts are being undertaken in specific priority areas, but could also generate challenges for ensuring co-ordination and coherence and exploiting relevant synergies. This is furthermore manifest in the varying ICT maturity across different policy areas.

The implementation structure of the Plan

The implementation of the Plan– like the previous information society Strategy– is co-ordinated centrally by the SETSI placed in the Ministry of Industry, Tourism and Trade. A broad, inclusive implementation approach was adopted following the positive experience of the first plan (OECD, 2010a). The implementation of the Plan thus involves all levels of Spanish public administration as well as industry and other society representatives, *e.g.* the consultative body, CATSI, which provides prioritisation input on the use of the Plan's resources.[1]

E-government initiatives are co-ordinated with the Ministry of Territorial Policy and Public Administration, responsible for e-government development in Spain, as well as with other relevant partners. The Ministry of Territorial Policy and Public Administration is also responsible for deploying and operating the shared services and interoperability infrastructure across all levels of government.

A public business organization, red.es, has been established under the aegis of the Ministry of Industry, Tourism and Trade to support the sharing and replication of best practices in key projects. Red.es is considered a highly professional trouble-shooting organization, engaging in high priority projects across different ministries and levels of government. Red.es can also provide assistance for initiatives undertaken outside the Plan on the basis of separate collaborative agreements. Red.es is primarily funded as a part of SETSI, but services can also be financed on a consultancy-like basis, further ensuring their quality.[2]

Additionally, numerous ministries and local authorities are responsible for projects within their own specific field of competence. One example is the modernisation programme in the Ministry of Justice, which is self-funded and run separately by the ministry – except for the few projects co-funded by the Plan. This underlines the non-exclusive character of Plan and its role in setting out an overall direction for future development.

The budget and financing mechanisms of the Plan

The previous information society strategy mobilised more than EUR 9 billion from 2006 to 2009 (OECD, 2010a). Given the general financial and fiscal constraints currently faced by Spain, no prior single amount is separately allocated to the Plan for the period covered. Instead, the budget is allocated one year ahead. This could be a way of enforcing strong continuous prioritisation, so ensuring that the goals of the information society strategy are also pursued within the ordinary budgets of the responsible authorities.

Table 2.1. **Budget dedicated to Plan Avanza 2, 2011**

Thousands EUR

(1000 EUR)	2010*	2011
Paperless administration	536 525	74 483
Infrastructure	–	124 924
Trust and security	7 147	233 256
ICT sector development	234 665	781 909
Total	778 337	1 214 572

* The 2010 budget reflects the initial division of Plan Avanza 2 into five strategic areas. This table reflects the later merger into four areas.

Source: Spanish Ministry of Industry, Tourism and Trade (2010), *Estrategia 2011-2015, Plan Avanza 2: Anexos*, Ministry of Industry, Tourism and Trade, Madrid.

Table 2.1 illustrates the budgets allocated among to the four strategic axes for the implementation of the information society strategy in 2011. The budget for overall e-government initiatives was around EUR 37.4 million in 2011 – approximately 3% of the planned annual budget. This does not include the budget for health and education covered by "paperless administration" in Table 2.1.

Loans are the most important financing source across all areas in 2011 – typically provided by the Official Credit Institute under the terms of an agreement with the Ministry of Industry, Tourism and Trade. However, this does not seem to apply for e-government. This might reflect the particular characteristics of investments in ICT infrastructure at large, as opposed to e-government projects with shorter depreciation periods.

The Plan's initiatives consist of cross-governmental initiatives as well as initiatives implemented within specific parts of national or regional government. This suggests that different parties are co-funding some initiatives. Co-funding is part of the implementation strategy and helps ensure that the organisations involved commit to the implementation of the projects.

The Spanish e-government strategy

E-government development has a strong legal basis in Spain: laws and formal regulation are considered key instruments in rethinking and implementing e-government services.

Spain has for the last five years built its e-government work on Law 11/2007 which governs citizens' electronic access to public services. It is now considering how to build on the progress achieved and address the challenges identified in order to ground future work. A key plank in that work, as the Plan outlines, is to make government paperless by 2015 (Box 2.1). The e-government setting is aligned with joint European goals set out in the Digital Agenda for Europe and the measures outlined in the European E-government Action Plan 2011-2015 (European Commission, 2010a; 2010b).

E-government legal framework: Law 11/2007 and the E-government Action Plan

In June 2007, the Spanish Parliament passed Law 11/2007 which enshrined people's right to communicate with public service administrations online (government of Spain, 2007a). The law has played a pivotal role in determining the e-government approach in Spain and in clearing the path for progressive work in this area. The end of 2009 was the deadline it set the public service for complying with its provisions, of which the most important entitle citizens:

- The guarantee of digital service provision, whereby public administration bodies should ensure that all government transactions and services are fully available and updated online.

- The right to choose among the service channels available when communicating with public authorities. The authorities are required to provide both analogue and digital communication and service processes as requested by citizens.

- The right not to supply data and documents already in the possession of other public administration bodies. Public authorities must organise data exchanges across all levels in formats that enable efficient interoperability.

- The right to secure and confidential storage of all personal data used in public authority files, applications and systems.

- The right to equality of access to public online services. government services shall not discriminate against citizens using non-electronic forms of communication and services.

- The right to obtain an electronic ID and to use other approved electronic signatures.

- The right to access personal data and files about ongoing processes.

Law 11/2007 was followed by additional laws and royal decrees widening its scope to encompass the promotion of digital communication and procedures across all areas of government.[3] The Spanish central government closely monitored and supported the implementation of the legislation in order to ensure it aligned with the four thrusts of the action plan to implement the Law 11/2007: citizen-oriented services, adapting administrative procedures to law, common infrastructure and services, and horizontal actions (government of Spain, 2007b).

The Law on E-government is applicable to national, regional, and local tiers of government. However, its implementation is limited by budgetary constraints and priorities at regional and local levels relating to local autonomy and the discretionary powers of local governments. There is therefore a somewhat lower degree of compliance at the regional level (Spanish Ministry of Territorial Policy and Public Administration, 2011).

Co-ordinating e-government

The e-government institutional framework is co-ordinated from the Ministry of Territorial Policy and Public Administration supported by the Directorate for the Promotion of Electronic Administration. The task of co-ordination at state level is the mandate of the High Council on E-Government which seeks to develop and implement action from a whole-of-government perspective, particularly on strategic projects. The focus is on getting all ministries and government bodies to realise the value of e-government in furtherance of paperless administration.

The council is a collegial body attached to the Ministry of Territorial Policy and Public Administration and chaired by the minister. It is responsible for setting the strategic direction on e-government, monitoring its implementation, and reviewing the e-government strategy plans of the individual ministries. It must approve all procurements above EUR 1 million – which covers hardware, software, and service agreements.

Other key members are the General Secretary for Public Sector Modernisation; the State Secretary for Telecommunications and Information Society; the State Secretary of the Treasury and Budget. Additional parties can be invited to the meetings if deemed necessary. The High Council meets once or twice a year and incorporates a technical support body, the Standing Committee of the High Council of E-Government. On the committee sit chief information officers or general directors from the different ministries who handle senior management co-ordination through monthly meetings. The Director General for the Promotion of E-Government chairs the Committee.

In addition to the national government co-ordination framework, cross-governmental bodies – which include sector and industry stakeholders from outside government – are also at work. The Sectorial Committee for E-Government, for example, is a technical body that ensures co-operation with central government, the autonomous communities, and local governments on e-government issues.

The e-government co-ordination structure seems to take as its point of departure input at national level – i.e. the funding of strategies and projects which are, if relevant, co-financed by regional or local authorities. The frequency and degree of co-ordination varies. The integration of the co-ordination mechanisms with the output side does not seem as well developed.

Figure 2.1. **Framework of institutional co-ordination**

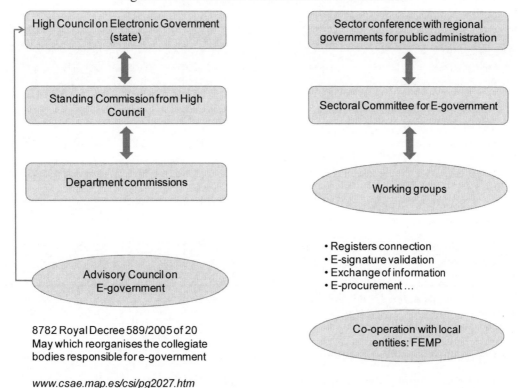

Source: Spanish Ministry of Territorial Policy and Public Administration (2011) "Executive summary of the report for the Council of Ministers of September 16, 2011, on the situation of e-government at the general public administration", Ministry of Territorial Policy and Public Administration, Madrid.

E-government budget and incentives

The budget for e-government is composed by national cross-governmental e-government efforts, ministries' own spending on IT and e-government within their own field of competency and, thirdly, regional and local authorities' spending which also forms an important part of the overall picture.

Solid data on e-government and IT expenditures has been collected at the national level for more than 20 years. Table 2.2 shows expenditures broken down into capital expenditures, human resources, and services.

Even though data on e-government expenditure are not collected consistently at regional level, it seems that regional e-government efforts are by far the greatest, both in budget size and in impact.

Table 2.2. **National ICT expenditures**

National government	2005	2006	2007	2008	2009	2010
ICT infrastructure						
ICT equipment (assets)	307 029	293 674	269 106	296 037	280 262	265 885
Expenditure on computer software	116 825	154 890	157 809	153 338	167 750	159 144
Human capital						
Number of individuals employed as ICT personnel	13 662	17 352	14 460	16 134	15 234	15 234
Public expenditure on compensation for ICT personnel	247 629	359 739	311 074	369 978	376 424	357 603
Public expenditure on ICT related training	5 112	4 580	1 748	3 973	4 970	4 715
Service capital						
Public expenditure on ICT services	429 795	474 398	486 247	575 227	684 342	649 235
Public expenditure on external ICT consultants	207 249	224 515	295 964	299 415	280 130	265 759
ICT equipment that does not qualify as investment	29 647	32 939	28 850	28 852	29 825	28 295
Total	1 343 286	1 544 735	1 550 798	1 726 820	1 823 703	1 730 636

Source: Spanish Ministry of Territorial Policy and Public Administration (2011) "Executive summary of the report for the Council of Ministers of September 16, 2011, on the situation of e-government at the general public administration", Ministry of Territorial Policy and Public Administration, Madrid.

Expenditure on ICT may be categorised under four different headings:

1. grants and/or loans within specific target areas (e.g. the funding that SETSI awards as part of the previous information society strategy);
2. loans for ICT investment facilitated by the Official Credit Institute and the lending authority (e.g. for investment in infrastructure targeted in the Plan),
3. co-financing agreements between national co-ordination ministries and the relevant authorities (e.g. digital justice projects financed jointly by the Ministry of Justice, regional governments, and separate the Plan's funds).
4. ministries' own ICT budget allotments, which enable direct project execution.

Although the Plan draws up a common set of e-government goals, the budget it allots to e-government initiatives seems insignificant compared with the overall national e-government budget.

Online service provision and the national channel strategy

As outlined above, Law 11/2007 sets public service administration demanding goals and standards in areas like online service delivery, internal procedures, and the handling of data. The results are clearly demonstrated in Spain's climb up international e-government ratings

Spain ranked in the top ten in a 2010 study into e-government maturity for the European Commission, which includes rankings for the availability and sophistication of 20 key public services. It came in eighth on both counts. In a further assessment, the User Experience of Websites and Portals, it was fourth (Lörincz et al., 2011).

Spain's online service provision today is of very high standard. The way it has developed is illustrated in Figure 2.2, indicating that around 90% of procedures (end-user services) and 99% of all (internal) administrative processes at national level are available in digital form. The figure shows, furthermore, how prioritisation fits into the Law 11/2007 implementation process, focusing on achieving high impact as fast as possible.

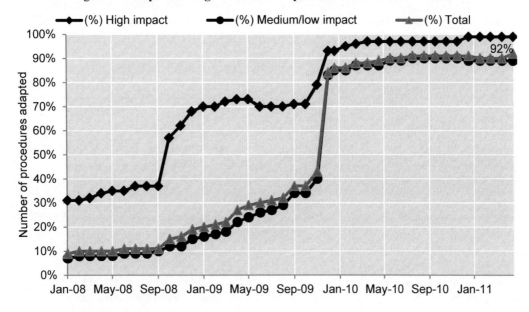

Figure 2.2. **Implementing online service provision under Law 11/2007**

Source: Spanish Ministry of Territorial Development and Public Administration, State Secretariat for Telecommunications and the Information Society (2011).

The most significant increase in Spain's online service provision took place in the second half of 2009, the deadline for complying with the law being the end of that year. The steep rise in the online service provision at national level was reflected in the regional implementation of Law 12/2007. Even though the law is not binding on local authorities if budget priorities suggest otherwise, it is estimated that they have made around 77% of all procedures available online (Fundación Orange, 2010). Measures aimed at implementing the law's more back-office-oriented requirements do not seem to have been factors in monitoring implementation.

Law 11/2007 provides an overall channel strategy which begins with a citizen's right to choose the channel through which to interact with the government. The government, in turn, should deliver services on the platforms that meet citizens' needs. The key components in the service channels currently used are:

- analogue service access using paper or through physical attendance;
- remote service access through automated voice services or call centres;
- Internet-based service access through "060.es" information portal or the "sede" websites of the relevant ministries;
- mobile-based service access through SMS or mobile application solutions.4

The extent of the authorities' precise service delivery obligations may be better understood in relation to the ICT maturity of the different users of specific services, the nature of the services, and the cost structures of the different access channels. However, specific cross-government priorities of service channels do not seem to have been formally established.

The service channels are tied together by joint infrastructure components. In this respect, a number of developments warrant special attention: the electronic national identity card (Box 2.2); the high-speed administrative network SARA, which connects all national and some local governments; payment solutions; standards to facilitate interoperability and security introduced by the Royal Decrees 3 and 4 in 2010 (OECD, 2010a).

Box 2.2. Spain's electronic national identity card

The Electronic National Identity Card (*DNIe*) is a mandatory smart card that is gradually replacing the former paper national identity card. Aside from its traditional functions, the new electronic card and its associated public key infrastructure (PKI) provide two electronic services:

- guarantee electronically a person's identity
- digitally sign electronic documents, giving them a legal validity equivalent to a handwritten signature.

This second feature is possible because the DNIe is a secure signature creation device (SSCD) and its microchip contains two qualified certificates: one for signing, the other for authentication. As a result, DNIe electronic signatures fulfil the requirements for legal validity set out in the EU directive on electronic signature.[1] It should be noted that this service takes advantage of technical and organisational synergies between electronic signature and electronic identity. By September 2011 more than 25 million electronic identity cards had been issued. Specific uptake across services and other signatures has not been observed. The image below illustrates the different physical components of the electronic National Identity Card.

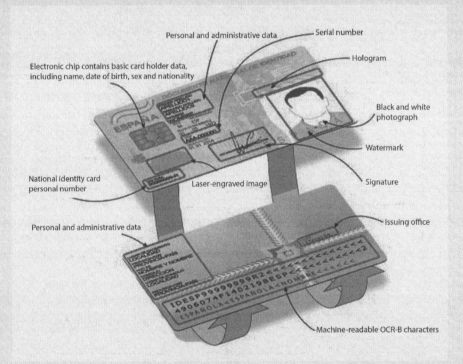

1. Directive 1999/93/EC of the European Parliament and of the Council of the European Union of 13 December 1999 on a Community framework for electronic signatures.

Source: Spanish Ministry of Interior (2012), *www.dnielectronico.es*.

However, the infrastructure underpinning the service channels and the ways they are used are still to be fully aligned across all levels of government. In 2010, all regional governments signed co-operation agreements entitling them to access common services and infrastructure (Spanish Ministry of Territorial Policy and Public Administration, 2011).

User uptake and the realisation of value

Mature online provision enables e-government to improve its impact and the benefits it may bring. However, actual uptake – the use of services – is often an important precondition for reaping the financial benefits.

The first measures of user uptake were collected by the E-government Observatory (OBSAE). Spain defines user uptake as the actual use of services relative to the total number of transactions. The uptake of e-government services also covers use through intermediaries, e.g. by delegation to social collaboration partners (Tax Agency, 2010b). Consequently, the way Spain counts uptake is different from the survey-based method of Eurostat, which measures the frequency of digital interactions with public service offices. Whereas OBSAE's numbers indicate that the civil service handles 51% of all transactions with citizens electronically and 82% with businesses, the survey based numbers sometimes lower than the actual transaction-based measures (Observatory of Electronic Government, 2011). However, these relate only to the national tier of government. The survey-based ranking can compare user uptakes internationally (Figure 2.3).

Figure 2.3. **Percentage of citizens using the Internet to interact with public authorities, 2005 and 2010**

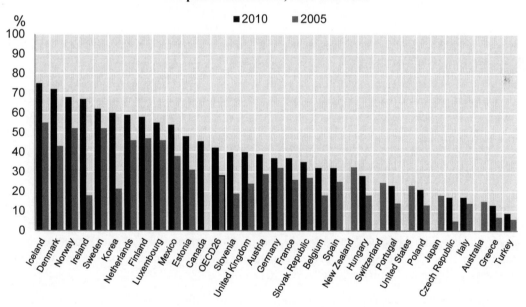

Source: OECD (2011), *Government at a Glance 2011*, OECD Publishing, Paris, http://dx.doi.org/10.1787/gov_glance-2011-en.

Eurostat data from 2010 show that although Spain had risen to practically 32% since 2005, it remained below OECD average of 42%. The picture changes in the EU context where Spain's level of user uptake measured in transactions is on a par with the EU's 27% average. Spain also lags behind its OECD peers in businesses' online dealings with the civil service, as Figure 2.4 shows.

Figure 2.4. **Percentage of businesses using the Internet to interact with public authorities, 2005 and 2010**

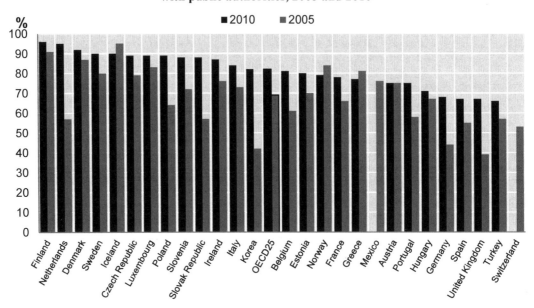

Source: OECD (2011), *Government at a Glance 2011*, OECD Publishing, Paris, http://dx.doi.org/10.1787/gov_glance-2011-en.

In 2010, 67% of Spanish businesses interacted with the civil service electronically against an OECD average of 82%. This trend is the same when Spain is considered in the EU context, particularly regarding transactional services (Eurostat, 2011). Spain's focus on the demand side and on increasing user uptake seems well grounded in the available data.

While almost all businesses with more than ten employees have an Internet connection, micro businesses still seem to lack behind. The same trend may be observed in the use of online services, where smaller businesses have far less dealings with the civil service than large and medium business. No comparable uptake data have been found on the 95% of Spanish businesses that have less than ten employees (Spanish Ministry of Industry, Tourism and Trade, 2011a).

Although the electronic ID is operational in all national and in many regional and local services, no aggregated data seem to exist on its specific use in online services. Though the service is also provided by private business, this is mainly related to banking. So although the overall majority of large businesses use the digital signature to communicate with public services, fewer seem to use it for communication purposes with their own clients or suppliers (Spanish Ministry of Industry, Tourism and Trade, 2011b).

> **Box 2.3. Key messages: E-government**
>
> - The economic and financial challenges faced by Europe have had a severe impact on Spain, forcing the government to take strong action to modernise the public sector and improve the efficiency of public service delivery. E-government can make an important contribution to this agenda.
>
> - Much of the current work on ICTs and e-government was conceived during years of economic growth in Spain and implemented during periods of recession or low growth. A focus on the financial gains and benefits of e-government still does not seem fully integrated in the guiding strategies and co-ordination mechanisms.
>
> - Law 11/2007 on citizens' electronic access to public services has resulted in significant e-government progress in online service provision. Spain seems now at the forefront internationally in online provision of mature services. Its efforts to improve the availability and quality of e-government services rank it high in the EU and UN e-government benchmarking.
>
> - However, the success parameters for these achievements seem to have focused on compliance with the law. Emphasis was placed on increasing the availability of services online to allow for online digital communication between the public service and the citizens. Key challenges now appear to be increasing e-government uptake of services and reaping the benefits of the investment in infrastructure and service provision. Spain seems already to have initiated important work on measuring e-government uptake in this regard.
>
> - The Plan covers e-government and sets directions within specific policy areas. However, it does not currently appear to be the primary driver of the strategic and actionable goals for overall e-government development. In order to reap the benefits of e-government, it will be crucial to ensure alignment between the Plan and any specific new policy instrument (e.g. strategies, action plans) that addresses e-government in the future.
>
> - Ensuring alignment between central government and regional and local government through joint governance and co-ordination mechanisms is essential to optimise the impact of e-government programmes. This might require the strengthening of the governance mechanisms both horizontally and vertically across all levels of government.

Digitising taxation, the early mover

Taxation is a complex area of public service upon which governments rely for collecting revenue. Because of the sheer complexity and number of transactions, there are great potential gains to be had from digitising procedures and processes. There is universal interest among OECD countries in using e-government to achieve greater efficiency. Thanks to its investment in digital tax administration, operations and services, Spain's National Taxation Agency now seems to have achieved cost-effective organisation as measured both in terms of annual ICT expenses and overall annual expenses relative to the taxes collected (OECD, 2011d).

E-taxation is considered an e-government champion and pioneer in Spain. With the first electronic exchanges of files and data dating back more than 20 years, the tax agency has a wealth of experience in digitizing its services.[5] It has intensified these efforts within the last years following the implementation of Law 11/2007.

Taxation is the responsibility of the Ministry of Economy and Finance. The National Tax Agency accounts for more than 27 000 of the ministry's 37 000 employees. It is a major player in the Spanish civil service with local structures in all 17 regions – most employees work at the local level. Some 1 000 employees are responsible for developing and operating electronic tax services, with the emphasis chiefly on development, but also on the service architecture. Core IT services are concentrated in-house.

Strategy and co-ordination

The National Taxation Agency's key duties include collecting national taxes like income tax, corporation tax and VAT; managing and supporting some aspects of regional tax collection in close collaboration with the autonomous regions; assisting in border management and the fight against smuggling; and enforcing the law as it relates to revenue collection.

Strategic planning

The Tax Agency engages in multi-year strategies in parallel to its annual planning. The current focus is on three overarching policies: the prevention of tax fraud; strategic communication; and the prevention and punishment of fiscal, labour, and social security fraud (Tax Agency, 2011b).

E-government is an important tool for addressing main priorities, such as the focus on greater interoperability and the exchange of data between government authorities. Following these strategic orientations, the Taxation Agency initiated a series of initiatives in 2011, which build on and enhance previous work on e-government.

The National Taxation Agency had built solid e-government foundations even prior to the development of the broader national initiatives of establishing e-government frameworks and adopting the subsequent regulations. It achieved significant results even before rolling out supportive national action. This might have contributed to the pragmatic approach adopted, which focuses on results and uptake rather than on specific infrastructural standards.[6] It is often observed that wide variations in e-government progress in certain policy areas make e-government co-ordination, synergies, and the rationalisation of systems more difficult. This challenge is shared to different extents by the most advanced OECD countries. It also seems equally relevant with regard to transnational co-ordination at the EU level, where different levels of e-government maturity make standard setting a challenging exercise.

Co-ordination mechanisms

The National Taxation Agency is represented in the High Council of E-Government as well as in the Standing Committee. As the taxation area can provide key data to be used in a broad range of government service areas, co-ordination remains important. The High Council's approval of the ministerial e-government plans together with the co-ordination rules on larger procurement approvals seem to some extent to ensure this.

Co-ordination with the broader information society policies, as put forward in the Plan, does currently not seem to be considered directly relevant for the Taxation Agency and they are not directly involved in or funded by the Plan's projects. No Plan funds have been applied to work on e-government in the taxation area. E-government co-ordination of taxation is essential, although it does not seem to emphasise national information society policies.

The structure of the National Taxation Agency reflects a high level of specialisation among a range of departments. For a better implementation of the new regulations on e-government, a task force with representatives from every service and department was created in 2008 within the Taxation Agency headed by the current Planning Service. This unit has been of key importance in ensuring necessary alignment, co-ordination and joint implementation; hence, the involvement of all departments and services in the Taxation Agency's e-government development and implementation. The technical work on e-government is primarily located in the Taxation Department for Information Technology. This department develops new projects and provides and manages new services. The department's board of directors is responsible for high-level co-ordination, management and the final internal review and approval of all projects.

The Taxation Agency has invested in e-government and today has annual IT expenditure that is below the OECD average.[7]

Digital service supply and channel strategy

All 426 tax-related digital services have been made available online in a fully transactional manner. They had handled 131 million cases as of January 2010 (government of Spain, 2010).

Channel strategy and approaches to service prioritisation

The taxation area provides services through the national portal, *060.es*, the Taxation Agency's own webpage which provides every electronic service with the highest security level, social media services, and telephone services, which include some text messaging. The agency also provides traditional paper-based service and person-to-person contact in its offices (Tax Agency, 2011a).

As well as internal government services and data exchanges, the overall strategy has been adapted to the two main external user segments: citizens and businesses. Although citizens can choose their preferred communications channel, some digital communications have in some areas been made compulsory for businesses. Most of the electronic tax return filings are already mandatory, and the Royal Decree 1 363/2010 furthermore recently mandated the Tax Agency to compulsory digital notifications of businesses. Though the decree's mandates are broad, the aim focuses mainly on companies and legal entities with guaranteed access (Tax Agency, 2010b). Hence, the main electronic communication point is the so-called "electronic office" (*sede electronic*) where citizens and businesses can access their personal accounts and files, or any notifications pending. The included notification service (for Authorised Electronic Access) is managed by the Postal Service and works somewhat similar to a webmail (Tax Agency, 2011b).

In line with the practices adopted by a large number of OECD countries, this approach underlines the focus on user uptake in order to realise the investment in the extensive online service provision.

Compliance with the Law 11/2007 was organised around a series of priorities, starting with high-impact services rather than those with low impact. The biggest annual tax declarations, measured by number of transactions, are shown in Table 2.3.

Table 2.3. **Key transactions handled by the Tax Agency**

Personal income tax (IRPF)	19 366 138
Corporation tax (impuesto sociedades)	1 405 275
VAT (IVA)	3 441 574
Excise duties (II.EE.)	9 332 650

Source: Tax Agency (2010).

These declarations illustrate only part of the potential of digital tax administration – additional quarterly or monthly returns are not covered here. Considerations on cost-intensity and digital readiness of the users of different digital services provided by the Tax Agency do not seem to have been fully developed. The current orientation of prioritising the online supply of services and access channels seem to have been only partially leveraged. Further exploring these areas might enable Spain to reap even higher benefits.

Tax system infrastructure and services

The Tax Agency relies on different types of infrastructure. These are, in principle, co-ordinated in accordance with the European and the national framework of service standards. However, specific priorities and different stages of development influence the development and choice of infrastructure and services in practice. One example is the use of electronic ID where several different types of electronic identifications are currently being used,[8] in line with the pragmatic approach adopted in the Tax Agency to advance the increase in user take-up. Another example is interoperability, where the national scheme sets the general standards to follow.

Digital service demand and e-government user uptake

The number of tax declarations submitted digitally has increased significantly in recent years, making the Tax Agency a leader in the national challenge to improve the user uptake of online services. The agency has developed experience that might be of broad value across Spain's public service sector and offers an opportunity to share good practices and experience acquired in the taxation area.

Progress in user uptake

The Tax Agency shows that 49.7% of the 19.3 million income tax returns in 2010 were filed remotely. This covers both the use of Internet services, telephone and text messaging. The variation in the use of remote solutions is relatively across the regions – from 61.5% in Murcia to 41.3% in Navarra (Tax Agency, 2010a). This moderate variation could be considered to be in line with the efforts of the previous information society strategy to ensure a harmonisation across the regions.

The Tax Agency measures the use of the online services provided. As in most OECD countries with advanced levels of e-government, however, there does not seem to be an established, comprehensive framework for measuring and advancing the user uptake at all levels of government. Though uptake data exist, they are not available across all provided service areas or user segments. This makes future decisions on improving user uptake more challenging.

Looking specifically at the online income tax returns, a significant and continuous increase can be observed over the last decade. This is illustrated in Figure 2.5.

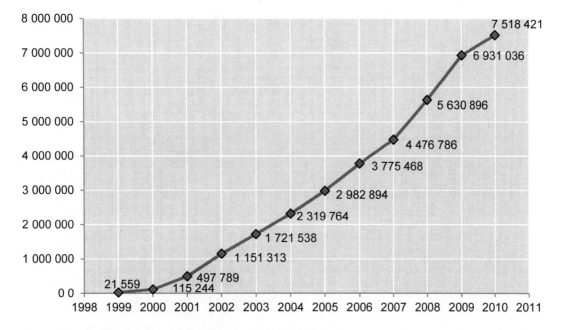

Figure 2.5. **Online income returns in 2010**

Source: Tax Agency (2011), *Memoria 201*, Tax Agency, Madrid.

However, the significant progress still reflects the fact that less than 40% of all personal income tax returns are submitted online. This is in line with the general online user uptake trend. It bears out Spain's current focus on the demand side and seems to indicate that the objectives of Spain on increasing user uptake is grounded and motivated by the available data on current uptake trends. Sixty-eight percent of total tax returns (in 103 different tax forms) are submitted electronically.

Marketing and communication

The Tax Agency rightly places emphasis on publicising the digital services it provides. It does so through its own specific initiatives and through the broader e-government national communication strategy and initiatives.

Spain is among the OECD countries that use social media technologies in tax administration, both internally as well as externally (OECD, 2011c). One example is the use of social media and video to give easily accessible information on how to use the electronic ID and the taxation services.[9] While it seems that social media is used mainly for presentations and communications purposes, additional potential applications are reportedly being explored. Figure 2.6 shows a public relations drive by the Spanish tax authorities to encourage people to file their income tax returns online.

Effective marketing and public relations can be key levers for improving user uptake. They help to raise awareness of the availability of online services and support the challenging process of cultural change. Existing networks – digital social media, but also social and private structures and communication channels – are also used. The already mentioned use of intermediaries in e-government uptake might be an indicator that the potential for using existing networks as an uptake accelerator still remains to be fully exploited.

Figure 2.6. **Social media marketing: taxation through YouTube**

Note: The figure shows a picture from a YouTube video by the Spanish Tax Agency, introducing the online service delivery channels. The video demonstrates how to get online and request the draft income tax file in order to process it. The video illustrates the process, beginning with a presentation of the electronic National Identity Card.

Source: Tax Agency (2011), video, *www.youtube.com*.

Consolidating and delivering returns on investment in e-government

The Tax Agency has set its sights on making tax administration efficient. However, it is difficult to establish clear measures of efficiency or the use IT in administration and service delivery. Data do exist on the cost of tax collection and the relative IT costs, though comparing them directly might produce ambiguous results.

Looking at the aggregate administrative costs of tax functions as a ratio of net revenue collections set out in Table 2.4, reveals a mixed trend. The improvement in this cost-efficiency ratio since 2011 might suggest a trend towards an increase in relative costs. Among many possible explanations is a likely one is that there has been a drop in collected revenues.

Looking at the period from 2005 in OECD countries, average IT costs in revenue bodies would seem to account for 11%-12% of the aggregate administrative costs (OECD, 2011d). Spanish IT costs relative to aggregate administrative functions seem very low at between 5% and 6%, which points to an effective use of IT as a result of the high online service provision and decrease in IT expenditure. This measurement does not, however, include IT personnel costs for Spain – inclusive data might blur the trend.

Table 2.4. **Trends in IT costs relative to aggregate administrative costs**

Taxation expenditures	2005	2006	2007	2008	2009	2010
Aggregate administrative costs for tax functions to net revenue collections	0.74	0.68	0.65	0.82	0.97	–
Total IT expenditures (thousands EUR)	108 527	100 629	110 943	99 651	84 976	95 916
IT costs to administrative costs for all functions	6.6%	6.8%	5.2%	6.1%	5.4%	–

Source: Presentation by Spanish Tax Agency (2011); OECD (2011d), *Tax Administration in OECD and Selected Non-OECD Countries*, Comparative Information Series 2010, Forum on Tax Administration, *www.oecd.org/tax/administration/47228941.pdf*.

In parallel, it seems that there is an increase in the share of salaries in the overall costs of tax administration (OECD, 2011d). This supports the need to ensure greater efficiency by leveraging the financial benefits of e-government to reduce the cost of human resources. Clear strategies for integrating and measuring personnel costs as part of the decision making and e-government project management have not been observed.

Project management model and business cases

The Tax Agency manages a very high number of ongoing projects. More than 700, differing in scope and size, are currently being managed using the same overall project implementation model that integrates combinations of scrum and lean thinking.[10] Approximately 95% of the projects are considered internal projects managed solely within the organisation.[11] This underlines the importance of internal governance and project management. The revenue department has developed a comprehensive dashboard of management information to monitor the progress of ongoing projects. This supports CASIR's decisions on e-government. Given the broader societal value of the data and projects developed in the taxation area, it seems important to ensure considerations on any broader governmental implications of digital services developed and provided by the Tax Agency, particularly in terms of interoperability.

Shared services

The Tax Agency operates all core business services internally, or "in house". This covers both physical hosting and maintenance of infrastructure (e.g. mainframes) but also the development of key systems and programmes. Efforts have been made by the Tax Agency to increase the standardisation and the reuse of code. This has resulted in several internal capacity-building programmes to professionalise the internal working processes. Internal strategic work on sourcing therefore exists and includes considerations on what services should be delivered in-house and what should be outsourced. This approach has not been observed across the government in accordance with Spain's general strategic guidelines on the digital service architecture. A more comprehensive whole-of-government strategy in this matter might enable further synergies, consolidations, and economies of scale.

> **Box 2.4. Key findings for taxation**
>
> - The Tax Agency is considered a frontrunner in e-government implementation. Investments made over a number of years have resulted in a high online service provision and in what seems to be an effective organisation.
>
> - A multi-channel approach establishing the choice of preferred service channels as a citizens' right seems well implemented. Some work on a stronger prioritisation of channels leading to higher efficiency has also been initiated in a few areas. Royal Decree 1 363/2010 mandates some compulsory digital communication in areas such as communication between businesses and the Tax Agency.
>
> - Shifting service channels can be a challenging process of cultural change for users. A public relations and marketing strategy to support smoother channel switches and increase uptake has been initiated and could be developed further.
>
> - The Tax Agency is using a commonly adapted internal IT project management framework, which particularly addresses the technical implementation of the initiatives. The business case aspects supporting decisions on costs, investments and benefits concerning e-government projects could be further integrated.
>
> - The Tax Agency manages a sizeable part of the required IT services and operations in-house. Focus on improving policies and achieving the benefits of e-government could also build on a plan for consolidating service provision and shared ICT infrastructure, also across levels of government.

Digital justice or modernising for a new social deal

The Spanish government has undertaken a process of modernisation of its justice system. "The Strategy Plan for Modernisation of the Justice System 2009-2012" has set in motion a series of projects to make changes to the justice system and is using digitization as a key lever to that effect. This section examines the approach adopted by the Spanish government in digitise its justice system. Aside from the overall strategy plan, modernisation plans also exist at the local level of the judicial system where a broad set of stakeholders are involved in the reform process. This parallel modernisation is co-ordinated partly under the Plan and partly under the national Strategy Plan for Modernisation.

Strategy and co-ordination to change the justice system

The Spanish Ministry of Justice has several responsibilities in such key areas as harmonisation and oversight of the law and the entire legal system. The Spanish judicial system comprises courts, tribunals and public prosecutors. They come under the responsibility of the Ministry of Justice, whose duty is to provide them with resources. Judges answer to the General Council of the Judiciary. The ministry also has responsibility for local courts and autonomous communities and is in charge of managing legal records and registries, including the Civil Register.

The Strategy Plan for Modernisation of the Justice System 2009-2012

The Strategy Plan for the Modernization of the Judicial System was a priority for the Spanish government in power from 2004 to 2011. Its political project was to make the administration of justice more open and transparent. It sought to bring it closer to ordinary citizens by simplifying its processes and procedures and improving its accessibility (Ministry of Justice, 2009). E-government was an important tool.

The strategy plan is grounded in political understanding of the need for a "social deal for justice" (*Acuerdo social por la justicia*) aimed at modifying the public image of a slow, inefficient judiciary.[12] The implication was that the outcome of the strategy should be measured both by the judicial system's improved internal effectiveness and efficiency and by how improvements were communicated to the public to improve the public perception.

The strategy plan is built on six pillars, or strategic axes, and structured to support its reform goals and programmes. Each strategic axis has a set of specific objectives, which include e-government-related programmes. Axis III is particularly relevant to e-government as its declared aim is to develop a technologically advanced justice system. Around half of the projects in the modernisation programme are directly IT-related, which makes Axis III particularly important. E-government and modernisation projects are considered complex to implement in the Spanish legal system, due to tradition, institutional inertia, and to the fact that some judicial institutions are autonomous and not under the ministry's aegis.[13] There was, therefore, a need for strong, close management and co-ordination and implementation efforts at all levels.

The Strategy Plan for the Modernization of the Judicial System in Spain sets an example of how, through e-government, to embrace and support the government's broader policy goals at all levels as an integrated part of a public-sector modernization project. Large-scale investments to improve organizations demand a high level of senior management skills.

Co-ordination structures and mechanisms

The Strategy Plan for the Modernization of the Judicial System in Spain comes under the responsibility of the Ministry of Justice. However, its implementation programme involves several other stakeholders, including the Ministry of Industry, Tourism and Trade and the Ministry of Territorial Policy and Public Administration. The autonomous regions also play an important role. The ministries and the regions parties co-finance the programme.[14] The projects that are part of the Plan's modernisation programme are subject to the broader co-ordination of red.es and SETSI.

Red.es and SETSI are involved only in the projects funded by the Plan, currently the programme Ius+reD, with the continuation of work on the Civil Registry also contributing to the broader modernisation of the judiciary (Ministry of Justice, 2011a).[15] The cross-governmental public-private hybrid working methods of red.es are regarded as highly beneficial to the national, and particularly the local, implementation of the IT projects on modernisation of justice.[16]

Further, the Ministry of Justice seems to have established an implementation model that takes the difficulty of organisational change into consideration, based on an understanding of the particular dynamics of the Spanish legal system.[17] The city of

Madrid was chosen as the starting point in the implementation and deployment process in order to demonstrate leadership and set a good example from the centre.

Budget for the modernisation programme

The cost of the Spanish court system appears higher than in European peer countries, although certain data may be ambiguous. Spain appears to have moved from a judicial system with expenditure around the EU average measured as a percentage of GDP in 2005, to being somewhat costlier than average in 2008-2009-2010 (Eurostat, 2011; CEPEJ, 2010).

The considerable budget increases indicate that the judicial system is a political priority. It might also reflect a rise in the work, i.e. in the number of cases handled. Whereas the government as a whole sought to reduce expenditure in 2011 by 16%, the Ministry of Justice went for a cut that was less than half that amount – 6.9%. Expenditure in the Ministry of Justice has increased by around 75% since 2004 (Spanish Ministry of Justice, 2010).

Figure 2.7. **The development of expenditure in the Ministry of Justice**

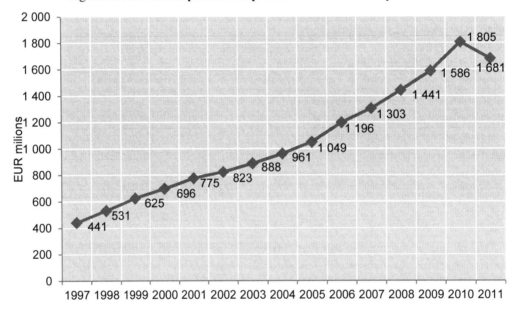

Source: Presentation on the *Proyecto de Presupuestos Generales del Etado 2011* from *www.mjusticia.gob.es*.

The high level of expenditures reflects a high level of investment during the same period. Funds allocated to the modernisation project for 2009-2012 exceeded EUR 600 million, with 2011 accounting for EUR 200 million (Ministry of Justice, 2010). The high level of investment increases the expectations for and the attention to the realisation of the benefits.

In 2011, funds for the modernisation programme were allocated across different parts of the programme, as illustrated in the figure above. Investment in IT (infrastructure, new technologies, systems, etc.) from Ministry of Justice funds accounted for the lion's share, from which it follows that the Plan projects and funds accounted for around 10% of the total modernisation budget in 2011. The numbers are indicative on the basis of 2011 only and do not cover the total funds or projects allocated to the modernisation programmes.

This might support the view that the main parts of the larger projects previously funded by the previous information society strategy, such as the digitisation of the civil registry, had been developed or even implemented at this stage. It may also indicate that the Plan currently does not play the most important role in the co-ordination, development and implementation of e-justice in Spain. The specific projects implemented, however, remain relevant, *e.g.* on digitisation of the courtrooms through audio and video conferencing facilities.

Figure 2.8. **Overall prioritisation of resources in the modernisation programme, EUR millions**

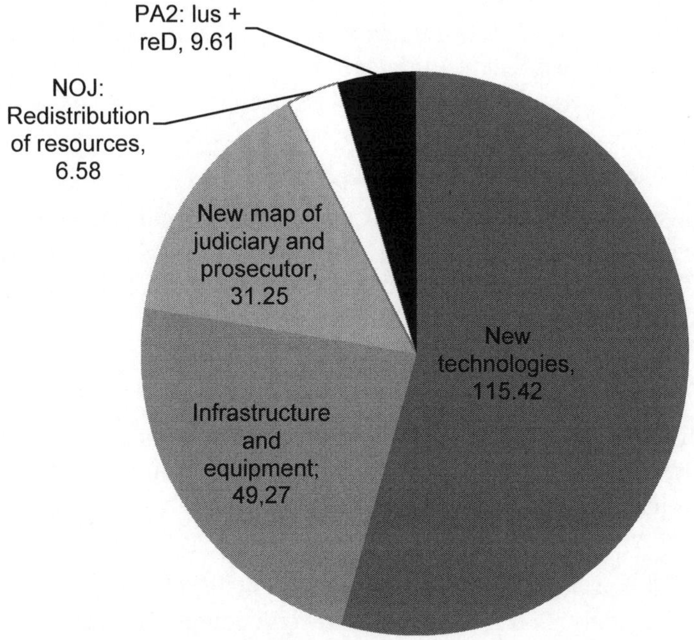

- PA2: Ius + reD, 9.61
- NOJ: Redistribution of resources, 6.58
- New map of judiciary and prosecutor, 31.25
- New technologies, 115.42
- Infrastructure and equipment, 49.27

Source: Presentation from *www.mjusticia.gob.es*.

Digital service supply and channel strategy

Law 11/2007 did not originally apply to the administration of justice in Spain, which in this regard can be considered "a late mover". Only services provided by the Ministry of Justice under the heading of general administration were covered. The large increase in online service provision from the Spanish public sector was not reflected in the administration of justice. Right after the enactment of the general e-government law, in January 2010, only 24 out of 76 services were available online, which accounted for around 40% of the annual transactions in the administration of justice (government of Spain, 2010).

In July 2011, the Law on Citizens' Rights to Electronic Access to Public Services was extended to the justice department under the terms of Law 18/2011. It coincided with the advance of the modernisation programme.

Service delivery portals and electronic office

The Ministry of Justice organises its online services offers according to the different user segments. Portals, for example, are therefore customised for citizens, prosecutors, civil servants, and court clerks. There is also a single portal for the administrative branch of the judicial system. The portals are currently at very different implementation stages and most remain to be fully developed. While the citizens' portal, for example, is still at an early stage, the administrative justice portal for the civil servants seems to be almost completed.[18]

Service delivery covers all channels, i.e. e-mail, telephone, mobile applications, text messaging, digital television, over the counter, and postal mail. And, in addition to the judicial system portals and central government portal "*060.es*", the different social media platforms are also increasingly a channel of service delivery.

The *Oficina Judicial* is considered a central component of the engine driving the modernisation process.[19] The *Oficina Judicial*, or "Judicial Office", denotes the instruments that support and enable the daily business of the judiciary – i.e. courts, judges, clerks, and the various administrative processes. Judicial offices communicate their decisions with legal professionals over the Lexnet, a secured two-way communication link.

One key digitisation challenge is how to fundamentally rethink the way that value is added in the processes of the organisations in order to fully exploit the transformative potential of ICTs. This remains a continuous imperative in work on e-government development and modernisation (OECD, 2009).

Transparency and digitisation

Until recently the judicial system had not experienced comprehensive digitisation of its services. As mentioned above, the implementation of some internal organisational changes seems to have advanced. However, key objectives remain to be achieved on the final part of the e-government axis in the modernisation programme.[20]

The new paperless procedures are being implemented in the courts where cases may be treated entirely digitally. Documents can be received digitally and if paper is received, it is digitised before it is handled. All documents created during proceedings are digital (Spanish Ministry of Justice, 2011c). Another element is the introduction of the eFidelius system, which replaces the written reports of the clerks to the court with video recordings of the court proceedings. The new digital nature of court proceedings requires greater internal flexibility and efficiency, but it gives judges, lawyers, administrative assistants and citizens' access to the same files and material online. Transparency has greatly increased and the workings of the Spanish judiciary have been transformed.[21]

The Ministry of Justice has also introduced a comprehensive system for internal transparency, monitoring the status of the case handling covering all levels of government, thus including also the progress in the user take-up of the new platforms. The management system, NOJ Minerva, reflects the processes included in the existing regulation, supports procedural management, and generates real-time data enabling closer follow up and management on key performance indicators. This creates a new kind of organisational transparency and a more detailed and accurate ground for management that is unlike any previous initiatives. Citizens can also access this information, which gives credence to the idea of a transparent justice system and may increase trust in public institutions.

Efforts to improve implementation and availability began at national level with national systems. They are now being extended to local level. National systems promoted by the Ministry of Justice, for example, are being made available for the autonomous communities. The Plan's project "Ius+Red" has so far resulted in the creation of 33 self-service kiosks for the online Civil Registry, the installation of 346 viewing and communication points in courtrooms, and the use of audio and video conferencing facilities to support the digitisation of the judiciary system enabling new services and more efficient processes (Spanish Ministry of Justice, 2011a).

It is not clear to what extent these initiatives can contribute to solving the more general challenge of increasing discontent and distrust of politics that has been expressed recently in Spanish society.[22]

Digital service demand and user uptake

The modernisation process in the Spanish Ministry of Justice is currently ongoing. However, even where legal, organisational and technical requirements have been put in place, it is probably too early to assess the results of implementation in terms of impact and uptake.

Marketing and communications

Communication skills are important for successful implementation and user uptake. Successful implementation of the modernisation process is also about transforming the image of the justice system. Communication thus plays an important role. In addition to traditional lines of communication, the Ministry of Justice has drawn up a web 2.0 communications strategy to convey the image of a modernised justice system to a wider, younger audience (Spanish Ministry of Justice, 2011b).

This involves presence on different social media platforms and channels as illustrated in Figure 2.9.

Figure 2.9. **Social media platforms and channels for communications**

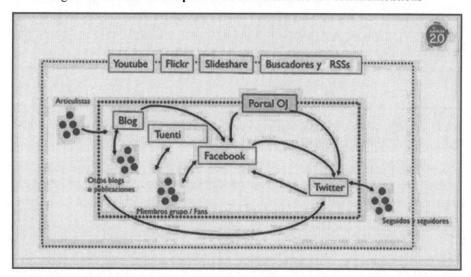

Note: The figure shows a picture from a YouTube video by the Spanish Ministry of Justice, presenting its use of social media for communication purposes. The picture shows the use of social media platforms (Twitter, Facebook, Tuenti, YouTube, Flickr, Slideshare), search engines, RSS feeds and blogs vis-à-vis the relationships between writers, content providers or linking sites, with group members, fans and followers. The Office of the Judiciary portal [Portal de Oficina Judicial] is the main focal point for Ministry of Justice's social media communication strategy.

Source: Spanish Ministry of Justice (2011), "Comunicación 2.0 del Ministerio de Justicia", Ministry of Justice, Madrid, *www.youtube.com/mjusticia?gl=US&hl=ca#p/a/672E5EF39206C901/0/_SpJ2c3eulA*.

So far the number of viewers and subscribers in particular seems limited.[23] This may be due to the fact that the initiative was launched only recently, in 2011. Changing working methods and shifting service channels can be a challenging process of cultural change. A public relations and marketing strategy to support the channel switches, increase awareness of the online service supply and increase uptake for the most efficient service channels might help to widen use.

User uptake and IT skills

Training programmes have been launched in support of the programme to modernise the judicial system. They are designed principally to help office staff as new digital procedures and systems are rolled out. The training drive includes systematic use of e-learning tools to promote implementation within the justice system.[24]

Considering that the digital justice strategy was launched recently, the primary focus still seems to be on developing and fully implementing the digitised services and systems to ensure a smooth organisational change (Spanish Ministry of Justice, 2011a). When implementation is completed, a shift in direction away from supply might be beneficial to support the higher user uptake reaping the rewards of investment.

Consolidating and realising investment

The Ministry of Justice has focused on project implementation and online service supply. It is now following up with an equally firm focus on completing implementation fully and assessing the achievements.

Realising investment and business cases

The modernisation of the Ministry of Justice was initiated by a strong explicit political demand.[25] It does not seem based on a specific business case, weighing the need for investments against the expected qualitative and financial gains. Accordingly, the implementation process has focused particularly on providing and managing the service delivery infrastructure.

Significant increases in expenditures – particularly the scope and level of IT investment in the last few years – are not always easy to manage. The needs in terms of IT project management capacity and skills can be difficult to determine and develop for projects as ambitious as the justice modernisation programme.

The financial – as well as the qualitative – benefits come from increased efficiency in processes, automation, and growing use of the established solutions. The Ministry of Justice is monitoring the uptake and progress. In order to quantify and benefit from gains, the ministry will need to consider how and when the resources will be "freed" from previous tasks. For example, should staff, made redundant by the new automated digital processes, be dismissed in order to reduce expenditure? Should the recruitment of new staff be stopped? If so, who would benefit from the gains of investment? Should staff be reassigned to other departments? If so, what competencies would they need?

Common infrastructure: the Civil Registry

The modernisation of the justice system is one major internal change project. But other projects contribute to the larger infrastructure of basic databases all shared by all the departments and offices in the public service. The Civil Registry is an example, originally initiated by the previous information society strategy and continued by the second.

Figure 2.10. **The three components of the judicial modernisation programme**

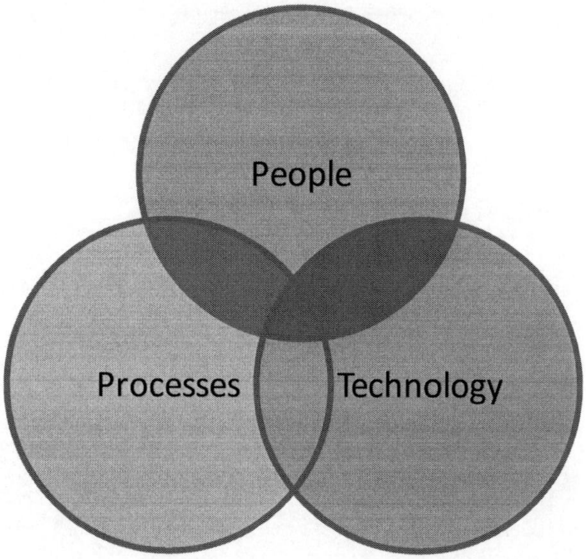

Source: With reference to: Spanish Ministry of Justice (2009).

The comprehensive reform of the Civil Registry in Spain is an important milestone in the modernisation of the Spanish justice system. The goal is to move from paper registers in local town halls to a single consolidated digital registry, accessible online through the application Inforeg. The Ministry of Justice has introduced a new organisational and technical model for the Civil Registry designed to achieve coherence between regulation, organisation and new technological opportunities (Spanish Ministry of Justice, 2011d). The register provides unique key data for all citizens and public authorities and includes the existing information from the old registries.

While it is widely recognised that the new Civil Registry provides improved access to important shared information within the Ministry of Justice, concerns have also been raised regarding the adaptation of back-office procedures – issues of privacy, optimal leverage of IT, cost reductions, higher efficiency, etc. (OECD, 2010c). Work on these processes continues through the implementation of the supporting programme "Ius+Red", both within and outside the Plan. Though increased efficiency through time savings are measured, there does not seem to be any evidence yet available of direct financial gains (Spanish Ministry of Justice, 2011e).

> **Box 2.5. Key findings for the digitisation of the justice system**
>
> - The Strategy Plan for the Modernization of the Judicial System in Spain is a good example of how to embrace and support the general political goals of government through an ambitious e-government programme that improves effectiveness and efficiency, trust and transparency. The considerable investment calls for a high level of senior management skills.
>
> - Services within the justice area are being transformed. Digitisation enables new processes but also poses new challenges for old administrative procedures. This transformation puts the emphasis on responding to users' demands. A priority area is indeed how to establish customised, user-centred services.
>
> - The modernisation of the judicial system has been guided, in particular, by the political priority of transforming the justice system to improve its efficiency and deliver trusted, higher quality services. Having focused on the service provision and met the timeline originally envisaged for implementing the main initiatives, it might be time for a stronger focus on realising the benefits of the significant investment in this area.
>
> - The Ministry of Justice has established a comprehensive, ambitious system for monitoring and managing the status of cases being handled and the progress in the user uptake, internally as well as externally. Such measurement might provide solid ground for realising investment and responding better to users' demands.
>
> - Changing working processes and shifting service channels can be a challenging process of cultural change. Training programmes have been launched to address these challenges. A public relations and marketing strategy to support the channel switches and increase the uptake of the most efficient service channels is important to ensure that expected results are achieved. New public relations efforts are also being developed using social media.

Examples of good practice in e-government

Good practices from the experience of OECD countries can be a rich source of inspiration for continuously improving national public administration. This section considers examples of good practice selected to demonstrate cases with relevance for the e-government challenges faced by the Spanish government.

The three sections below highlight good practices in strategic management, channel strategy and focus on user uptake as well as models of agile organisation. Unless stated otherwise, cases are based on data requested directly from government departments and received by the end of 2011.

Strategic tools for management and alignment

The ICT business case model: renewing Danish ICT governance

Following the recommendations of a government working group in May 2010, the Danish government decided to implement a series of measures to increase professionalism in the management and implementation of larger ICT projects (Danish Digitisation Agency, 2011).

The background was recognition of the need to deal with an increasing number of large ICT projects characterised by overspending on budgets, exceeded timelines, and lack of delivery on anticipated solutions. The government adopted the recommendations, also inspired by examples from private businesses, and took a new approach in three key areas:

- the use of common methodologies and stronger competencies;
- focus on managing high-risk ICT projects; and
- a better co-operation with and evaluation of government suppliers and advisers.

The use of common methodologies includes the systematic measurement of ICT project management competencies. The government has introduced the use of one project model for ICT projects for all central government bodies, advisers, and suppliers. To complement the common project methodology, mandatory use of a common business case model (Box 2.4) is required for all ICT projects exceeding DKK 10 million (approximately EUR 1.35 million) to clarify the value added of the proposed project. A plan for realisation of investment should also be submitted along with the business case.

Box 2.6. The focus of the business case model

- Promoting the understanding and ease of cross-functional transfers.
- Standardising input (assumptions) and output (key figures), thus giving decision makers a more comprehensive background for taking decisions.
- Realistic budgeting providing the foundation for improved project and risk management.
- Clarification of financial risks (along with the classic risk analysis) through the business model.
- Benefit realisation and costs where the financial controllers are central players in the design, follow-up, and realisation of business cases.

In addition, a new division – the Government Project Division – was established to provide advice and assistance from senior resources applying the new methodology. Formal certification of ICT project leaders was also introduced to promote project management as a career path.

Furthermore, a strengthened governance setup was introduced. It includes, among other key measures to better manage high-risk projects, the establishment of the Government Project Council, which performs systematic quantitative risk assessment that takes into account funding, scope and consequences, resources and competencies, technology and ownership focusing on the realisation of the benefits. The systematic risk assessments are supplemented by external reviews carried out on the recommendations from the Government Project Council.

Further to this, five strategic principles were formulated to support the management, development and implementation of ICT projects, committing the Danish government:

- not to be "first mover" on the use of immature technologies;
- to reuse already purchased or developed solutions;

- to implement only projects with clear benefits relative to the costs;
- to divide projects into smaller, simpler, more easily implementable parts; and
- to ensure project management based on common methods and highly professional competencies.

Denmark is currently implementing the recommendations and the Project Council and a new Government Project Division were established within the Ministry of Finance in 2010.[26]

Measuring e-government maturity – the Mexican case

Mexico faces, as do most OECD countries, the challenge of determining how well ICT resources are being managed and whether high expenditure actually delivers value. The Mexican Government therefore established a model to quantify the maturity of e-government projects in terms of their value. This move enables it to centrally assess the overall maturity and level of the services delivered, so laying the ground for determining where to get the most value for money (Mexican Ministry of Public Service, 2011).

The e-Business Value and Maturity Model is an important addition to ICT management tools and an important addition to the federal government's management of its ICT resources.

The overall goal of the new maturity model is to contribute to ensuring:

- the capacity of government institutions to deliver public goods and services efficiently;
- citizens' ease of access to public goods and services, so reducing transaction costs;
- the facilitation of the infrastructure needed in the information society.

These overall goals translated into the model's more specific objectives: assess public value for citizens; assess the development of ICT projects, processes and services; integrate ICT-leveraged information to produce an infrastructure that enables higher efficiency and better services; optimise ICT expenditure and exploit synergies between interdependent institutions; and support government departments in ICT use and management and exploit the operational efficiencies.

The model uses three categories of indicators as Figure 2.11 illustrates.

The model allows the benchmarking of all federal government institutions identifying areas where significant improvements can be made on the basis of data on strategic programmes for more than 200 government institutions. This laid solid ground for effective digital governance. Good practices can also be highlighted and deployed in other government units.[27]

Figure 2.11. **Maturity measurement for e-government**

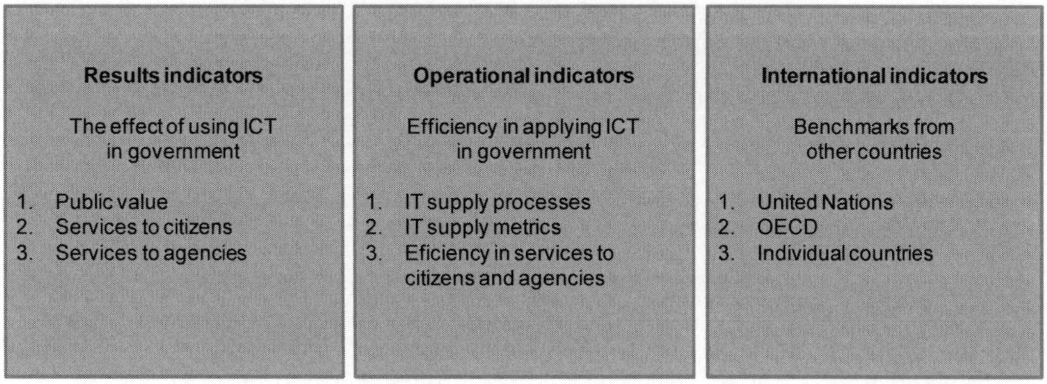

Results indicators	Operational indicators	International indicators
The effect of using ICT in government	Efficiency in applying ICT in government	Benchmarks from other countries
1. Public value
2. Services to citizens
3. Services to agencies	1. IT supply processes
2. IT supply metrics
3. Eficiency in services to citizens and agencies	1. United Nations
2. OECD
3. Individual countries

Source: OECD (2011), *Towards More Effective and Dynamic Public Management in Mexico*, OECD Public Governance Reviews, OECD Publishing, Paris, , *http://dx.doi.org/10.1787/9789264116238-en*.

Increasing uptake – managing channels and personalising services

Uptake by means of incentives in the taxation area – the case of the United Kingdom

Her Majesty's Revenue and Customs (HMRC) seems to have had great success in driving both its individual and business customers to online services. HRMC provides over 250 services online, receiving over 70 million returns and 98 million visits to the website each year. Individual customers filing self-assessment returns online reached an all-time high of 78% in 2010. Business customers now file their corporation tax, pay-as-you-earn tax, and VAT returns online. HMRC sees the increased availability and use of digital services as vital in achieving its objectives of reducing costs, improving customer experience, and increasing tax yield (Cabinet Office, 2011a).

HMRC has a wealth of experience and expertise in opening up access to its systems and data to third parties, while safeguarding the confidentiality of its customer information. This collaboration works to provide customers with a choice over how they interact with HMRC, as well as providing a service to the customer at virtually no cost to HMRC.

The success seems to have been driven by a mix of incentives that include, on one side, later deadlines for online filing and generally providing a more convenient service to customers. On the other side, mandating drives near 100% take-up of a digital service, but comes with greater enforcement costs and a sub-optimal customer experience.

HMRC has developed a set of principles to guide the future design and delivery of digital services and drive their usage on policy areas (Cabinet Office, 2011b):

- Delivering services "digital by default", but never striving to be an exclusively digital organisation. This means placing digital at the core of a mix of channels providing customers offering.

- Designing services around customers and driven by insight into their needs and behaviours. This means primarily using incentives to drive usage and uptake.

- Providing customers with the easiest and quickest service digitally will drive and retain high levels of uptake. This means continuing to open up access to systems and services for use by intermediaries.

- Providing permanent assistance to those customers who will always need help using the digital services and will signpost it to them accordingly. This recognises that there will always be customers unable to use the digital services for a variety of reasons, and supports them in fulfilling their obligations. This could be, for example, through the use of another channel or an intermediary.

- Focusing on delivering digital services that allow interaction with customers online, rather than simply transact with them. Continuing to ensure the digital channel is secure and trusted. This recognises that the security of the services is of upmost importance to ensuring the continued success of the digital services.

- As we move forward, looking to use digital services to increase transparency of customers' tax affairs. This will enable them to view many of their different liabilities and services, and drive a greater understanding of their individual tax liability.

This approach aligns with the Government Digital Strategy. It defines a "digital by default" user-centric approach. It focuses on understanding customers' needs and behaviours and on continually collecting feedback and information about the way they use the services to drive the improvement and development of them.[28]

Awareness campaigns to increase uptake and ensure alignment – eDays in Denmark

The Danish e-government strategy has applied the marketing concept of "eDays" in all of its previous national e-government programmes since 2001. The overall idea is to set up a series of e-government goals, related to citizens' rights, use of joint infrastructure components, or fulfilment of national requirements. Thus eDay is the day when new ways of using ICTs in government must be implemented (OECD, 2010b).

This implies, on one side, co-ordinated efforts in effectively informing users – citizens and businesses – of the new opportunities and conditions for communicating with government. This could be through advertising in the print press, on television, on the Internet, or in the social media. On the other side, it involves aligning government administration the goals to be met and implementing the necessary ICT solutions and adjusted processes in their organisation. In that sense, eDays merge e-government implementation and public relations. The concept of eDays is considered to be of great value for implementing e-government in the Danish public sector and increasing user uptake of the implemented solutions.

The third eDay was launched on 1 November 2009 to promote and implement the use of shared infrastructure, particularly for citizens. More specifically, it set goals designed to further easy, secure access to all digital services:

- "All national digital self-service solutions with a need for secure identification must use the common EasyLogOn service with the common digital signature. This contributed also to promoting the national citizen portal, borger.dk, providing a single-sign-on to the digital entry point to the public sector across all levels of government.

- All citizens can get a digital Document Mailbox through which they can send and receive secure mail to and from the public authorities. This means that citizens can chose to receive and lead all correspondence with the public sector in their Document Mailbox. This mailbox also covers digital communication with a series of private businesses using the same ICT solution.

- All public authorities can be contacted through the mailbox, and all the authorities can send all the relevant mails (individually or mass mails) and receipts through the Document Mailbox." (Danish Ministry of Finance, 2010)

The Danish Ministry of Finance (2011) has announced a fourth eDay in its recently published Danish e-government strategy, *The Digital Path to Future Welfare*.[29]

Co-ordination across levels of government – SOA in Germany

Co-ordination and alignment of service delivery between institutions and across levels of government is facilitated through the Administrative e-Service Directory (*Deutsches Verwaltungsdiensteverzeichnis* – DVDV) in Germany.

The DVDV lists electronically available e-government services and meets an important need for creating a secure, reliable communication infrastructure based exclusively on open Internet protocols and allowing cross-organisational, paperless processes. In operation since January 2007, it has helped more than 5 200 German civil registration agencies to save more than EUR 1 million per month.

Worldwide, it is one of the first and largest standardised service-oriented architecture (SOA) implementations in the government area. It was made possible through unique co-operation between various levels of government and sectors in the federal republic of Germany. The DVDV's range of applicability is not limited to civil registration but is open to any kind of communication with and between public administrative bodies in Germany. Such e-governance communication is either government-to-business (G2B) or government-to-government (G2G). Besides civil registration communication, the DVDV also supports processes in areas of government like tax administration and the judicial system.

At the Lisbon Ministerial Conference 2007, the DVDV won the e-Government Award 2007 in the category "Effective and efficient administration".[30]

Searchability to access public services and increase user uptake: gob.mx

The Mexican government has re-conceptualised the citizens' portal, *gob.mx*, in order to increase the uptake and availability of government digital services delivery, while supporting the public sector's use of social media and cloud computing. The key word is "search" – building on a partnership with the private company Google – as illustrated in Figure 2.12.

The increased accessibility and comparability between the different government institutions and services also leads to improvements in the internal management which provide incentives to ensure that services, access to information, and stronger citizen participation are fully exploited and optimised.

The portal's search engine is designed to be integrated with searches and the content of all government institutions in order to maximise the government investment in technology. Further, the portal facilitates the design and development of simple applications to be published directly on the site, which also makes it a "container" of applications and automated procedures concentrated in one place where citizens have access through an advanced electronic signature.

Figure 2.12. **Citizens' access to government services, Mexico**

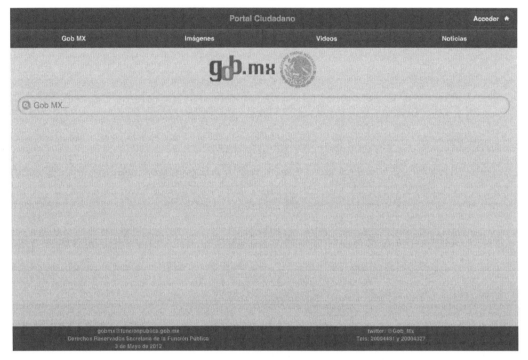

Note. The figure shows the Mexican federal government portal created in partnership with Google. The search portal approach helps organise citizens' access to government services while providing the government with important information about citizen needs and search patterns. This portal offers a search service specialised in the Mexican government and a personalised page for accessing public procedures and services.

Source: www.gob.mx

The portal uses advanced searches as well as social networking technologies as part of its leveraging of ICTs to improve and simplify citizens' access to government services. The portal includes the introduction of open standards for developing social networks and government applications to public services and processes. This also ensures a consistent framework for the use of ICTs in the public administration so that citizens perceive the government as a consistent whole.

The analysis of data on the use of services by the citizens enables the government to monitor usage preferences and prioritise service digitisation. This can support a strategy for service delivery channels that emphasises building digitisation on the demand of services already accessed online.[31]

Organising for agility

Partnerships to enhance service delivery channels – use of private networks in Italy

Reti Amiche (Friendly Networks) is a network of delivery channels in the private sector that gives citizens easier access to public services in Italy. This is an innovative project from the Ministry for Public Administration and Innovation. It aims to develop more pervasive and efficient interaction between citizens and government through the collaboration of private networks with public administration services. This initiative is expected to further facilitate relations between citizens and government and minimise delays in the provision of services and eliminate queues. The final goal is to relieve public offices of user congestion and allow more time and resources for new services (OECD, 2011e).

Meeting this goal entails involving existing service providers for public service delivery, such as post offices, tobacco shops, banks, pharmacies, police stations, train stations and distribution centres (malls) to simplify service access, diminish service delivery time, ensure friendly service and reduce the digital divide. This virtuous circle will increase customer attraction to private networks (tobacco shops, malls, railway ticketing web services, ATMs, etc.), and at the same time supply access to public services and information.

The reduced transaction costs brought about by ICTs has increased the number of interactions between customers and private companies or public offices, and has also increased the security levels of networks and the capacity to protect privacy, detecting and sanctioning illegitimate access to relevant information. This means that today it is possible, and tomorrow it will be easy, to link public service delivery to private networks, multiplying the channels for direct interaction between citizens and government.

The ultimate goal is to overcome inefficiency, lack of commitment and inadequate attention to customer satisfaction, which to some extent has characterised monopolistic delivery of the service by government administrative bodies. Instead, there will be an alternative to the classic public office queue, supplied by private delivery channels in shops, through web services and via ATMs. Friendly Networks compel government offices to give customers the attention they would get from private commercial businesses, reducing the degree of monopolistic concentration of public service delivery.

The increase in the number of access points will facilitate the usability of public services. The most innovative characteristic of Friendly Networks is the inter-operability between public administration web services and private networks. The aim of the Malmö Declaration is to put individuals, with their needs and demands, at the centre of e-government services. The delivery of personalised transactions, with the support of the one-stop-shop model, enhances the accessibility of the services by citizens. Agreements have been put in place to multiply physical points of access to services (from 60 000 to 100 000 by the end of 2010).[32]

Integrating Criminal Justice in Korea

Korea has been working on digitising its Criminal Justice System since 2005 (government of Korea, 2012). More than USD 75 million has been invested through the following steps:

- 2005: Informatisation strategy.

- 2005-2006: Case investigation system and internal portal construction for the police.

- 2006-2007: Decision system for prosecutors' investigations and trials.

- 2007-2008: Execution support system for the Ministry of Justice.

- 2009: Infrastructure (H/W, S/W).

Building on this work, paperless criminal justice has been achieved by creating integrated digital criminal justice processes including filing cases, forwarding, decision-making, ruling, and executing. This integrated system was first applied to the area of drunk/unlicensed driving cases in January 2010.

The purpose was to establish a "one-stop shop" for the provision of citizen-focused services, in order to avoid visits to government offices. The time for processing the approximately 340,000 ~ 540,000 annual cases has now been reduced from 120 days to 3–4 days. The increase in work efficiency is estimated to result in annual savings of USD 150 million. Furthermore, redundant information can be reduced and errors in paper-based administration prevented.

Open platforms to leverage transparency in Chile

An essential aspect of the Chilean Law on Access to Public Information has been the incorporation of electronic mechanisms for both proactive publication of information from government agencies (termed "active transparency"), and the management of citizen consultations on public information (termed "passive transparency") (government of Chile, 2011).

In this context, the General Secretariat of the Presidency developed an online system that simplifies the proactive publication of transparency information from public institutions using open standards that ensure a consistent user experience. At the same time a centralised portal was created to facilitate the search for information from different agencies: *gobiernotransparentechile.cl*. Regarding passive transparency, the government developed a free and open source solution that allows public agencies to manage information in response to requests made by citizens. Administration agencies must report electronically statistics of citizen requests into a system called the "Transparency Observatory" that tracks the delivery of transparency requests at a whole government level.

Several additional cases that promote fiscal transparency from different angles are worth mentioning:

Mercadopublico.cl, the open e-procurement platform. Since a new Law on Public Procurement came into force in 2003, *mercadopublico.cl* has expanded rapidly. The platform was created to increase the transparency of government purchasing decisions by publishing all public acquisitions through an open centralised portal. This initiative reduces transaction costs and widens the net of suppliers for government agencies. It has also allowed more small and medium-sized enterprises (SMEs) to participate in the vast public procurement market.

Analiza.cl, a business intelligence platform that seeks to deliver better information for better business decisions. It provides consolidated information and detailed bids and purchase orders that are traded on *mercadoublico.cl*. This single platform integrates tools that make it easy to find and view information from all government agencies. This platform also allows users to perform their own mash-ups based on procurement data and, for example, to build geo-referenced maps of public procurement.

Dipres.gob.cl provides open access to budget reports. In Chile, the use of technology to promote fiscal transparency is encouraged during budget formulation and execution processes. One of the main measures is the timely publication of budget execution reports of public agencies. Reports from all government bodies are made available to the public monthly and quarterly through *dipres.gob.cl*.[33]

Key assessments and proposals for action

Spain faces a series of serious challenges related to its difficult economic context. The global and European financial and economic crises have also hit Spain. Low growth, unemployment rates above 20%, and severe fiscal imbalance and budget challenges in the aftermath of the economic crisis are key issues faced by the newly elected government. Instability and rising easing interest rates in the euro area reinforce the need to ensure trust in Spanish economic policies and government administration. Spain has already undertaken some far-reaching measures, such as public expenditure cuts, reductions in public sector wages and employment, and reduced investment in public infrastructure.

The new Spanish government has clearly reinforced the need to put in motion the necessary measures to address these challenges. In particular, the Prime Minister has reiterated the importance of using modern information and communication technologies to sustain efficient and effective public administration at all levels of government, as this may foster increases in the overall competitiveness of the national economy. This is why all future measures are expected to support a reform of administration that eliminates inefficiencies and duplications, and sustains the overall aims of downsizing, restructuring, and reduced operating costs in the public sector.

The new government has restated the importance of carrying on the necessary reforms of the justice administration, which is not only seen as a power that guarantees protection of citizens' rights, but also as an essential factor in national competitiveness and trust in institutions. Ensuring that the administration of justice is a modern and quality public service is therefore seen as a priority. It is envisaged that, in the years to come, this should be achieved by promoting the efficient, co-ordinated use of new technologies.

E-government can provide part of the answer by enabling higher efficiency and effectiveness in government administration and service delivery. This can, in turn, support the government's ambitions to improve the budget in ways that might seem more acceptable to the electorate than direct cuts on welfare services. As such, the strategic use of e-government in the current context is imperative in order to increase public sector innovation, agility and mobility, and to enhance public sector productivity and secure sustainable public sector reforms. This was also reflected in the discussion of the latest session of the OECD Public Governance Committee in the fall of 2011.

This study has examined the Spanish goal to achieve a paperless administration by 2015 as a part of its information society strategy. It has focused particularly on an overall analysis and assessment of the digitisation of the areas of taxation and justice and it has proposed future action that is relevant to the broader e-government development within the Spanish public administration in general.

The section below summarises the key findings of the analysis on e-government, e-taxation and e-justice in line with the scope of this study and makes proposals for action to support the future efforts of the Spanish government in those areas.

Strategic alignment and strong co-ordination

A clear e-government strategy setting out a strong vision to be made operational through selected key actions can help to ensure a common direction for and alignment of individual e-government efforts. The strategic direction can be supported by the adequate co-ordination structure, proper incentives, a clear division of responsibilities, and follow-up of the progress in implementation through monitoring and evaluation through the use, for example, of indicators.

Key assessments

The Plan is a comprehensive information society strategy. It covers a broad range of policies to support economic recovery and growth. E-government is an important axis in this strategy, which sets the ambition of achieving a paperless administration by the year 2015. The work on e-government in recent years has been guided by the Law 11/2007 on citizens' rights to electronic access to public services. The significant increase in the online service provision indicates that the work to ensure this right seems to have been largely accomplished.

The Plan is based on a multi-stakeholder approach to co-ordination and implementation. The co-ordination measures established within the Plan could provide an adequate framework for fostering collaboration in funding the implementation of information society policies and related initiatives. However, the co-ordination framework could be further strengthened with regard to the overall e-government implementation.

Additionally, the overall economic challenges faced by the government could be reflected and addressed more strongly in the current e-government orientations, while the overall economic goals of the information society strategy could establish closer links to the more specific covered initiatives related to e-government. An exhaustive e-government co-ordination framework might go beyond the scope of the Plan, which covers and funds important e-government initiatives in some areas (e.g. health and education), but not all (e.g. taxation). This further accentuates the importance of defining an e-government strategy and the co-ordination of its implementation, particularly by the High Council of E-government and the Ministry of Territorial Policy and Public Administration. This seems in line with current efforts and considerations on how to advance a new e-government strategy for the years to come. E-government development and implementation challenges seem to be particularly important at local levels of government. The existing governance framework for e-government does not seem to account fully for this.

The strategies and governance frameworks vary within the different policy areas. The Ministry of Justice has put forward an ambitious, comprehensive strategy for modernisation. It provides a good example of how to embrace and support the general political goals of the government at all levels through e-government as an integrated part of public sector reform. The strategy to accelerate the process of digitising the judicial system seems to be guided by the political need to ensure public trust and confidence through a new "social deal". This has lead to significant investment in the modernisation process guided mainly by the political imperative of transparency and trust rather than by an underlying business case. The Spanish justice system seems to be still in the process of completing the organisational changes and reaping the dividends of transformation achieved through significant investment. This requires an ongoing high level of senior management attention. Further, ensuring and building trust in the Spanish justice might also be facilitated by establishing a high level of engagement and participation of the users, i.e. the citizens and businesses.

The strategy for the digitisation of the taxation area has focused, rather, on increasing revenues and efficiency. For example, it has sought to reduce fraud and increase user uptake of online services. The Tax Agency is considered a front-runner in e-government and large-scale investment seem to have resulted in a high online presence, mature channel priorities, and more effective organisation. Finally, the ICT infrastructure in the taxation area appear well grounded in, and co-ordinated with, the national governance framework for e-government. However, some co-ordination challenges still seem to remain in this regard, particularly in the use of electronic ID and interoperability.

Proposals for future action

Align e-government policies with public sector reform goals

Given the budgetary and financial challenges the Spanish government is facing, it could consider a closer alignment of its next e-government strategy not only with government reform policies recently announced by the new Prime Minister to specifically address budgetary and fiscal goals and challenges, but also with the objectives of administrative simplification to improve business competitiveness. E-government policies and strategies should embrace and support the general goals of the government at all levels, including the autonomous communities. Using ICTs to improve management by creating transparency and benchmarking of regional and local service provision seems an important approach to develop further – the lessons from the justice modernisation programme should be learnt in this matter. The importance of the economic challenges and fiscal imbalances could be key levers to using e-government in support of the ambitious government transformation process. This could help to increase overall government efficiency and secure the realisation of the financial benefits of e-government initiatives.

Strengthen the governance framework

Whereas the Plan sets out some of the overall directions regarding e-government, Spain could consider separately developing a specific e-government strategy. This would support the development of focused policies and the setting of concrete e-government goals. It would also facilitate e-government co-ordination across and within levels of government. The lofty Spanish ambitions for e-government might benefit considerably from a stronger institutional and governance framework. This may ensure alignment of goals and efforts both at the national as well as at the regional and local levels. It would involve not only revising the existing institutional framework but also the frequency, intensity, and the scope of existing co-ordination mechanisms as well as the central-local division of responsibilities.

Comprehensively ground the next e-government strategy

Building on this study the Spanish government could consider conducting a comprehensive e-government review in order to solidly ground the redefinition of policies and the development of a new e-government strategy. One area of attention could be to focus on the use of ICTs to improve the Spanish business climate, national competitiveness, and innovation. Given the importance of the Spanish regional and local levels of government, the review of the roles, responsibilities and capacities of the respective authorities should be covered by such a review. Assessing and measuring e-government performance at the regional and local levels would help Spain strengthen a whole-of-government perspective and improve multi-level co-ordination within and across levels of government. This would enable it to strengthen its position at the forefront of OECD countries in the development and implementation of e-government. Additionally, the review could help identify concrete action to improve the use of ICTs in support of administrative simplification designed to foster businesses innovation. One example would be to put in place a single window to access the public sector and thus eliminate the need to use different points of access to the public sector for each phase in a product's life cycle.

Prioritising digital service supply and service delivery channels

A channel strategy for service delivery is an important policy tool for managing and prioritising government interaction with citizens and businesses. An overall channel strategy encompasses analogue and digital entry and delivery points, and sets priorities between them.

Key considerations in the prioritisation of service delivery channels could be transaction types and frequencies, the readiness of user segments, the cost structures of the transactions on different channels, and the differences in service levels perceived by the users. Service delivery channels might be managed by combinations of push and pull mechanisms ("sticks and carrots") that range from regulating for mandatory use on the one hand to delivering user-oriented digital services of the highest quality on the other. Granting users the right to choose their preferred service channels, while providing incentives to use the prioritised digital channels can be designed in different ways.

Key assessments

Spain has made significant progress in online service provision. As a result, the country now performs above the average level of its peers in the OECD and the EU in this regard, according to EU E-government Online Availability measures: Spain stands at 95% against an average of 82% in 2010 for the 27 EU member countries. Further to this, Spain ranks among the top ten countries in the UN E-government Survey 2010: it is ranked ninth by the E-government Development Index and third by the E-Participation Index.

Law 11/2007 recognises the right of citizens to choose their preferred communication channel, thus laying the ground for a multi-channel strategy. Even though this is not translated directly into cross-governmental prioritisation of channels, some overall indicative estimates of the cost structure of different service delivery channels have been established in order to help ground and motivate any prioritisation process.

The government prioritisation services which should be fully digitised will require high political and administrative attention. Such work should reflect key policy goals at the same time as transaction intensity and the costs associated with the different service delivery channels. Spain seems well aware of the challenge of establishing a strong service delivery channel strategy that takes users' preferences and demands into consideration.

The Tax Agency has used a multi-channel approach to prioritise a selection of channels in the taxation area. This is leading to higher efficiency through the closing down of analogue channels. One good example is the mandatory use of electronic communication between businesses and the Tax Agency. Another example is the focus on automating communication by, for example, moving away from mail-based communication.

As for the judicial system, the strategy for providing services online reflects the Ministry of Justice's policy goals and modernisation objectives, as outlined above. The newly established digital platforms in the justice system (e.g. the NOJ Minerva management system and the digital Civil Registry) aim for ambitious, unprecedented levels of transparency and more effective internal working processes.

Proposals for future action

Prioritise online channels more strongly

The high provision of online services in Spain is a great advance in terms of the quality of public service delivery. However, to fully exploit the benefits of the considerable investments made in its online service provision, Spain could consider strengthening the use of the most efficient service delivery channels, e.g. automated services or online services. This could include further strengthening the obligation for certain user groups to communicate through selected service channels and providing incentives for the use of online services – rather than relying on a guiding principle of letting users choose the preferred communication channel. This would be in line with the practices of other OECD countries. Under considerable pressure to identify savings, several OECD governments are exploring how advances in technology-driven modernisation offer opportunities to generate significant efficiency and productivity gains while maintaining vital frontline services. A case in point is the strategic approach of the United Kingdom in making public services digital by default and delivering efficient, cost-effective public services that are responsive to the needs of citizens and businesses.

Supply better data to ground channel priorities

Good data to support the elaboration and reinforcement of a channel strategy are essential. Spain could consider establishing a framework providing for indicators and data related to the different service delivery channels that could support sound priorities of government policies on digitisation. Examples could be data related to differences in service delivery cost structures, user preferences, and services demand. Such a measurement framework should be clearly aligned with the overall e-government strategy, including the service delivery channel strategy. This would subsequently need to be incorporated into local strategies as long as local conditions were taken into consideration. Assessing the impact of ICTs on the overall delivery of policies and services would entail collecting relevant data from all levels of government. This would support optimising the use of ICTs across the whole government and contribute to the improvement of the overall performance of the public sector. This approach would be in line with the strategic orientations expressed by the members of the OECD Network on E-government at the Workshop on E-government Indicators in December 2011.

Simplify e-government access and focus on user value

Spain could consider how to further strengthen the user orientation of government service provision. This could involve reorganising services around users through digitisation and fundamentally challenging traditional organisational culture within the administration, regulation and working processes to focus on increasing user value. The high level of online service provision, as well as the developed e-government infrastructure, provides a good foundation for strengthened efforts to ensure seamless service delivery. A stronger focus on the simplification of rules, administration and service delivery, as well as on the value added for users across all levels of government, could additionally contribute to further reducing administrative burdens for citizens and businesses. One important element in addressing this challenge might be to engage the different users more strongly and more systematically in the development and simplification of the online services.

Increasing user uptake

User take-up of e-government services is a key measure of the degree to which the e-government services provision is actually used. The concept of user take-up refers to the use of public service delivered through new channels (i.e. adoption of e-government services). Together with survey-based consideration, data on user take-up can provide valuable contributions to policy making. One example of survey-based data includes Eurostat's uptake measures on online interaction between governments and citizens. More specific data on user take-up can be identified in relation to the overall number of transactions of the delivered service. Spain has strengthened its focus on the uptake of online services relative to the total number of a certain type of transactions. Additionally, Spain also extends the notion of e-government uptake to include the uptake of e-government services through intermediaries, e.g. by delegation to social collaboration partners.

User take-up of e-government services usually depends on a broad range of factors; and its increase can be addressed through several approaches; e.g. simplification, situation-bound customisation, engagement, marketing, and channel strategies.

Key assessments

Spain seems to have recognised the challenges of increasing user uptake of e-government services, particularly in light of its accomplishments in the increase of its online service provision. This also seems grounded in data as Spain, despite its high level of online service provision, remains just around or below the EU and OECD average levels of e-government user uptake – for both citizens and businesses. In 2011, for example, 32% of the Spaniards used the Internet for interacting with public authorities, against an EU average of 41% and an OECD average of 42%. Where 69% of the businesses in the EU's 27 member countries used the Internet for transactional services with the public sector, the figure for Spain was 65%. National initiatives to address this challenge are being developed. Frontrunners like the Tax Agency might provide inspiration in this regard – electronic declarations of corporate tax and VAT are among the highest in the EU.

Progress in the overall improvement of e-government service uptake seems to derive particularly from the good results in uptake of online services in the taxation area. The extensive digital service provision, in particular the actual user uptake of services delivered online, is essential for high efficiency in the taxation area. Currently, almost 40% of personal income tax declarations are submitted online. However, as a share of the total number of tax returns, 68% are submitted electronically.

The judicial system does not yet appear to have progressed as much or as far as taxation with regard to user uptake, particularly as its services were only recently digitised. However, the implementation of restructuring process seems well advanced with several deep-reaching changes having been made. One example is the digitisation of court proceedings, where case handling has become paperless through the use of the case handling system REGIUS. The Ministry of Justice has further established a comprehensive and ambitious system for monitoring the status of cases being handled at all levels of government. This also includes progresses in user take-up.

With regard to the raising of users' awareness of the availability of online services, both the justice system and taxation areas seem to be covered by ambitious communication strategies supported by the use of new technologies that include social media platforms. The impact achieved through these channels is not yet clear, although the initiative does demonstrate the large steps taken by Spain to progress in the "government 2.0 experience" by leverage, for example, YouTube channels and Facebook. The awareness of online services seems likely to be furthered through social networks.

Proposals for future action

Adopt common take-up measurements to ground policies

Solid data are a prerequisite for good policy making. As highlighted in the previous set of proposed actions, Spain could ground and support its work to increase online service user by drawing up a common framework for assessing the delivery and uptake of services through the various channels at all levels of government. This could support increases in user uptake and further enable management by realising Spain's overall investment in e-government efforts as well as in specific projects.

Use marketing to increase awareness

Having achieved a high level of online service provision, the challenge now facing Spain is to ensure an optimal level of user take-up. Spain could consider strengthening and targeting its communication and marketing strategy for e-government services. Such a strategy might also benefit from the leverage of existing social networks (analogue as well as digital) within government and through private business or civil society when deemed relevant. Clear prioritisations on how to achieve the highest impact is a prerequisite for success.

Exploiting and enhancing IT competencies

Skills, competencies, and trust in digital government are important prerequisites for harvesting the gains of e-government. Spain could consider addressing the need to improve the IT-related skills of businesses and citizens through designated capacity building and training strategies. Furthermore, these efforts could be considered in relation to existing cultural, educational and social policies and should seek to take advantage of existing initiatives and networks – particularly the related objectives of the Plan.

Focus on consolidation and return of investment

A consistent focus on producing returns on investment in e-government involves exploiting its financial potential along with the qualitative benefits. All e-government policies should therefore be aligned to ensure coherence, reduce redundancies, and promote mutual synergies. The professionalisation of project management in the public sector and consolidation of IT and e-government services are key components in securing return on investment.

Key assessments

A mature e-government strategy builds on a clear architecture of public services defining the roles and responsibilities of all stakeholders as well as the standards to link them. However, wide differences in e-government maturity and development within government can make co-ordinating initiatives and efforts difficult. It might lead to redundant infrastructure components – e.g. digital signature services and the under-exploitation of core data in the administration – and thus hinder the development of synergies and achievement of benefits. To tackle this matter, Spain has outlined a common government service architecture and seems to provide essential shared services – the SARA network, payment solutions, the e-delivery system for exchanging information between government administrative bodies and the electronic signature, DNIe.

The Tax Agency appears advanced in terms of its internal IT organisation. It is using a professional project management framework to guide the technical development and implementation of IT projects. A business case model to support decisions on costs, investment, and benefits of e-government projects could be drawn up and adopted to complement this framework.

The Tax Agency manages a certain amount of its IT services and operations in house. The sourcing strategy seems based on what the agency considers vital internal key competencies. Important in-house resources are currently focused on IT software development and service-oriented architecture.

Even though the transformation of the justice system is also guided by efficiency considerations as part of the new "social deal justice", the focus on measuring and realising dividends could be even further accentuated. This would be in line with the ongoing transition in the modernisation programme, moving from implementation towards achievement of results, i.e. towards more effective, efficient daily operations.

Proposals for future action

Use business cases to focus on benefits realisation

To support a stronger focus on the financial benefits of e-government, the Spanish government could consider introducing the use of a common project management model, which would include a business case tool that specified the costs and benefits of key projects. Experience in OECD countries has demonstrated that business case models can be useful in order to assess the value (benefits, costs and investment) of e-government projects. In addition to ensuring better decision making, the advantage of using business cases is that they can help to specify the value added of a project while enabling a clear focus during the implementation process on how to realise this value. The adoption of standardised IT project management and business case models would be in line with the practice in OECD member countries such as Denmark.

Standardise infrastructure and common components

The Spanish progress on e-government reflects concerted, co-ordinated efforts across government. To support this, Spain could consider clarifying even further the government service architecture to spell out clearly responsibilities across all levels of government – within each ministerial department as well as at regional and local levels. This could support co-ordination and collaboration efforts; provide guidelines on when who should apply what (e.g. on standards and joint components); and help strike the balance between the perspective of a single organisation as opposed to broader national perspectives.

Consolidate and pursue economies of scale

Spain could consider establishing a plan for the joint consolidation of ICT infrastructure services. As digital services evolve across government it is important to ensure the overall coherence and realise potential economies of scale, e.g. in management, operations, and sourcing. A consolidation of ICT infrastructure could be designed within government administration (e.g. through shared services) or through private service suppliers (e.g. subcontractors or partnerships), or in combination, reflecting the strategic decisions on what competencies to maintain and/or build inside public administration. The use of cloud computing – i.e. IT services provided on demand through the Internet – is increasingly considered by OECD countries to enhance government agility and performance.

Notes

1. El Consejo Asesor de Telecomunicaciones y Sociedad de la Información (Spain's Telecommunications and Information Society Advisory Board [CATSI]) was established by Royal Decree 1 029/2002. See OECD (2010) for a detailed description.

2. Red.es and information provided during the interview mission.

3. A more comprehensive view of the e-government regulation can be found in the Spanish Ministry of Territorial Policy and Public Administration (2010).

4. A more comprehensive view can be found at *http://administracionelectronica.gob.es*.

5. Interview with the director of the Tax Agency, 2011.

6. Interview with the director of the Tax Agency, 2011.

7. See OECD (2011d). However, the figures given are indicative and not directly comparable.

8. Interview with the director of the Tax Agency, 2011.

9. See for example the YouTube channel, *www.youtube.com/user/agenciatributaria?ob=5*.

10. Scrum refers to a software development model based on an agile, incremental development strategy. Lean refers to a management philosophy based on different parameters, particularly the focus on value creation, employee engagement and reduction of waste throughout processes.

11. Interview with the director of the Tax Agency.

12. The high priority of reforming the justice system was first launched with the inauguration speech of the previous government's President, see Spanish Ministry of Justice (2009).

13. Interview with the director of the Programme for Modernisation of the Ministry of Justice.

14. Ministry of Territorial Policy and Public Administration excepted, see Spanish Ministry of Justice (2009).

15. However, exemptions from this practice exist: red.es manages a project on the modernisation of justice in Galicia with a budget around EUR 2 million.

16. See OECD (2010a) for a detailed assessment of the efficiency of the inter-institutional co-ordination in Plan Avanza.

17. Interview with the director of the Programme for Modernisation of the Ministry of Justice.

18. Spanish Ministry of Justice (2011a). While the citizens' justice portal is 24% completed the administrative justice portal for civil servants is 95% completed, both

as of October 2011. It should be noted that the percentages do not reflect any strategic or outcome-based measures but an average of achieved milestones only.

19. See *http://oficinajudicial.justicia.es* for elaboration.

20. The goals within the third axis on a technologically advanced justice system are estimated to be 76% completed (Spanish Ministry of Justice, 2011a).

21. See, for example, YouTube presentations from the Ministry of Justice (2011)

22. The social movements that centre around the protest grouping, *los indignados* (the indignant), could be seen as an expression of this mistrust and discontent.

23. The official YouTube channel, *www.youtube.com/user/Mjusticia*, has 189 subscribers. The channel has been viewed around 55 000 times (as of December 2011). The number of times each video has been viewed appears limited – from more than 5 000 to just a few hundred.

24. A dedicated e-learning programme has been set up in response to the need to boost IT skills and introduce new ways of working.

25. See the presidential inauguration speech mentioned in Spanish Ministry of Justice (2009).

26. Additional information can be found at *www.digst.dk/Statens-projektmodel*.

27. Additional information on the government maturity model is available at *www.cidge.gob.mx/doc/ConsejoEjecutivo_130411.pdf*.

28. Additional information can be found at *www.cabinetoffice.gov.uk*.

29. Additional information on the new e-government strategy and related initiatives can be found at: *www.digst.dk/Digitaliseringsstrategi/Download-strategien*.

30. Source: *www.epractice.eu/cases/dvdv*, accessed 4 October 2008.

31. Additional information can be found on the portal, *www.gob.mx*.

32. Additional information can be found at *www.funzionepubblica.it/lazione-del-ministro/servizi-per-il-cittadino/reti-amiche/presentazione.aspx*.

33. Additional information can be found at *http://gobiernotransparentechile.cl*.

Bibliography

Cabinet Office (2011a), *Government ICT Strategy*, Cabinet Office, London.

Cabinet Office (2011b), "The role of ICT in delivering major policy initiatives", practices collected through the OECD Network of E-Government Officials, 16 December 2011.

CEPEJ (2010) (European Commission for the Efficiency of Justice), *European judicial systems Edition 2010 (data 2008): Efficiency and quality of justice.* European Council Publishing, Strasbourg.

Danish Digitisation Agency, "Business Case Model", *www.digst.dk/Statens-projektmodel*, accessed 17 February 2012

Danish Digitisation Agency (2011), "Overview of selected Danish e-government good practice cases 2011", practices collected through the OECD Network of E-Government Officials, 2 January 2012.

Danish Ministry of Finance (2010), "Status on Danish e-government efforts", unpublished background report to the OECD E-Government Review of Denmark, ..

Danish Ministry of Finance, Local Government Denmark (LGDK) and the Danish Regions (2011), *The Digital Path to Future Welfare, The eGovernment Strategy 2011-2015*, December.

epractice.eu, German Administration Services Directory (DVDV), e-practice case page on DVDV, *www.epractice.eu/cases/dvdv*, accessed 17 February 2012.

epractice.eu (2012), "eInclusion factsheet – Italy – areas (2012)", "Inclusive eGovernment Status on Inclusive eGovernment", last edited 16 January, *www.epractice.eu/en/document/329690*, accessed 17 February 2012.

European Commission (2010a), *A Digital Agenda for Europe*, European Commission, Brussels.

European Commission (2010b), *The European eGovernment Action Plan 2011-2015*, European Commission, Brussels.

Eurostat (2011), *Information Society Database*, European Commission, Brussels.

Fundación Orange (2010), "eEspaña 2010", Fundación Orange, Madrid, *http://fundacionorange.es/fundacionorange/analisis/eespana/e_espana10.html*.

Government of Chile (2011), "E-government: good practices", practices collected through the OECD Network of E-Government Officials, 17 December 2011.

Government of Korea (2012), *Integrated Justice System*, practices collected through the OECD Network of E-Government Officials, December 2011.

Government of Spain (2007a), Ley de Acceso Electrónico de los Ciudadanos a los Servicios Públicos, Government of Spain, Madrid.

Government of Spain (2007b), *Ley de Acceso Electrónico de los Ciudadanos a los Servicios Públicos, Plan de Actuación*, Government of Spain, Madrid.

Government of Spain (2010), *Informe Sobre el Complimiento de los Compromisos definidos en la Ley de Acceso Electrónico de los Ciudadanos a los Servicios Públicos*, Government of Spain, Madrid.

Her Majesty's Revenue and Customs (HRMC) (2011), "Digital by default", consultation document, HRMC, London, *http://customs.hmrc.gov.uk/channelsPortalWebApp/down loadFile?contentID=HMCE_PROD1_031509*.

Lörincz, B. et al. (2011), "Digitizing public services in Europe: putting ambition into action – the 9th benchmark measurement", report prepared by Capgemini, International Development Corporation (IDC), Rand Europe, Sogeti and Danish Technological Institute (DTI) for the Directorate General Information Society and Media of the European Commission, European Commission, Brussels.

Mexican Government (n.d.), "gob.mx", citizens' portal to e-government services, *www.gob.mx*, accessed 17 February 2012.

Mexican Ministry of Public Service (2011), "OCDE – Mejores Practicas Mexico", practices collected through the OECD Network of E-Government Officials, 15 December 2011.

Observatory of Electronic Government (2011), "Dossier of e-government indicators December 2011", Spanish Ministry of Territorial Policy and Public Administration, Madrid.

OECD (2009), *Rethinking public sector services: citizen-based approaches*, OECD Publishing, Paris, *http://dx.doi.org/10.1787/9789264059412-en*.

OECD (2010a), *Better Regulation in Europe: Spain 2010*, OECD Publishing, Paris, *http://dx.doi.org/10.1787/9789264095076-en*.

OECD (2010b), *Denmark: Efficient e-Government for Smarter Public Service Delivery*, OECD e-Government Studies, OECD Publishing, Paris, *http://dx.doi.org/10.1787/9789264087118-en*.

OECD (2010c), *Good Governance for Digital Policies: How to Get the Most Out of ICT: The Case of Spain's Plan Avanza*, OECD Information Society Reviews, OECD Publishing, Paris, *http://dx.doi.org/10.1787/9789264031104-en*.

OECD (2010d), *OECD Economic Surveys: Spain 2010*, OECD Publishing, Paris, *http://dx.doi.org/10.1787/eco_surveys-esp-2010-en*.

OECD (2010e), *OECD Information Technology Outlook 2010*, OECD Publishing, Paris, *http://dx.doi.org/10.1787/it_outlook-2010-en*.

OECD (2010f), "Survey Report: Survey of Trends and Developments in the Use of Electronic Services for Taxpayer Service Delivery, Forum on Tax Administration's Taxpayer Services Sub-group", OECD, Paris.

OECD (2011a), *Government at a Glance 2011*, OECD Publishing, Paris, *http://dx.doi.org/10.1787/gov_glance-2011-en*.

OECD (2011b), OECD Key ICT Indicators, *www.oecd.org/sti/ICTindicators*.

OECD (2011c), "Social media technologies and tax administration", Information Note, Forum on Tax Administration's Taxpayer Services Sub-group, OECD, Paris.

OECD (2011d), "Tax administration in OECD and selected non-OECD countries", Comparative Information Series 2010, Forum on Tax Administration, *www.oecd.org/tax/administration/47228941.pdf.*

OECD (2011e), *Towards More Effective and Dynamic Public Management in Mexico*, OECD Public Governance Reviews, OECD Publishing, Paris, *http://dx.doi.org/10.1787/9789264116238-en.*

Spanish Ministry of Industry, Tourism and Trade (2010a), *Estrategia 2011-2015, Plan Avanza 2*, Ministry of Industry, Tourism and Trade, Madrid.

Spanish Ministry of Industry, Tourism and Trade (2010b), *Estrategia 2011-2015, Plan Avanza 2: Anexos*, Ministry of Industry, Tourism and Trade, Madrid.

Spanish Ministry of Industry, Tourism and Trade (2011a), *Tecnologías de la Información y las Comunicaciones en las PYMES y grandes empresas españolas*, Ministry of Industry, Tourism and Trade, Madrid.

Spanish Ministry of Industry, Tourism and Trade (2011b), unpublished working paper on the electronic national ID.

Spanish Ministry of Interior (2012), *www.dnielectronico.es.*

Spanish Ministry of Justice (2009), *Plan Estratégico de Modernización del Sistema de Justicia 2009-2012*, Ministry of Justice, Madrid.

Spanish Ministry of Justice (2010), "Los Presupuestos para el Ministerio de Justicia, Proyecto de Presupuestos Generales del Etado 2011", presentation, *www.mjusticia.gob.es.*

Spanish Ministry of Justice (2011a), "Compromisos 2011", October 2011 monitoring report, Ministry of Justice, Madrid.

Spanish Ministry of Justice (2011b), "Comunicación 2.0 del Ministerio de Justicia", Ministry of Justice, Madrid, *www.youtube.com/mjusticia?gl=US&hl=ca#p/a/672E5E F39206C901/0/_SpJ2c3eulA.*

Spanish Ministry of Justice (2011c), "Expediente judicial electrónico en la Audiencia Nacional", Ministry of Justice, Madrid, *www.youtube.com/mjusticia?gl=US&hl=ca#p /a/f/0/zsO_IAyZ1hQ.*

Spanish Ministry of Justice (2011d), "Informes de Modernización Judicial en España: El Registro de Civil de Servicios", Ministry of Justice, Madrid.

Spanish Ministry of Justice (2011e), "Registro Civil: un antes y un después", Ministry of Justice, Madrid.

Spanish Ministry of Territorial Policy and Public Administration (2010), "Administración Electrónica, Textos Legales", *Boletín Official del Etado*, Madrid.

Spanish Ministry of Territorial Policy and Public Administration (2011), "Executive summary of the report for the Council of Ministers of September 16, 2011, on the situation of e-government at the general public administration", Ministry of Territorial Policy and Public Administration, Madrid.

Tax Agency (2010a), "Campaña de Renta 2010 Income Tax information campaign", Tax Agency, Madrid.

Tax Agency (2011a), *Memoria 201*, Tax Agency, Madrid.

Tax Agency (2011b), "Planificación de la Tax Agency 2011", presentation, Tax Agency, Madrid.

United Nations (2010), *E-Government Survey: Leveraging e-government at a Time of Financial and Economic Crisis*, United Nations E-Government Surveys, United Nations, New York.

ORGANISATION FOR ECONOMIC CO-OPERATION AND DEVELOPMENT

The OECD is a unique forum where governments work together to address the economic, social and environmental challenges of globalisation. The OECD is also at the forefront of efforts to understand and to help governments respond to new developments and concerns, such as corporate governance, the information economy and the challenges of an ageing population. The Organisation provides a setting where governments can compare policy experiences, seek answers to common problems, identify good practice and work to co-ordinate domestic and international policies.

The OECD member countries are: Australia, Austria, Belgium, Canada, Chile, the Czech Republic, Denmark, Estonia, Finland, France, Germany, Greece, Hungary, Iceland, Ireland, Israel, Italy, Japan, Korea, Luxembourg, Mexico, the Netherlands, New Zealand, Norway, Poland, Portugal, the Slovak Republic, Slovenia, Spain, Sweden, Switzerland, Turkey, the United Kingdom and the United States. The European Union takes part in the work of the OECD.

OECD Publishing disseminates widely the results of the Organisation's statistics gathering and research on economic, social and environmental issues, as well as the conventions, guidelines and standards agreed by its members.

OECD PUBLISHING, 2, rue André-Pascal, 75775 PARIS CEDEX 16
(42 2012 08 1 P) ISBN 978-92-64-11060-1 – No. 60143 2013-01

Printed in Great Britain
by Amazon